ALL WE CAN SAVE

Truth, Courage, and Solutions for the Climate Crisis

Edited by Ayana Elizabeth Johnson & Katharine K.

Praise for

ALL WE CAN SAVE

"A mosaic that honors the complexity of the climate crisis like few, if any, books on the topic have done yet . . . a feast of ideas and perspectives, setting a big table for the climate movement, declaring all are welcome."

—*Rolling Stone*

"A powerful read that fills one with, dare I say . . . hope?"

—*The New York Times*

"A groundbreaking collection of essays and poems by women at the forefront of the climate crisis movement . . . It's a sorely needed glimmer of hope—a reminder that there is a way out of this mess: collective action. . . . To quote the recently departed Ruth Bader Ginsburg, 'Fight for the things you care about, but do it in a way that will lead others to join you.' *All We Can Save* does that. Consider it required reading."

—*Ms.*

"A fascinating new anthology."

—Bill McKibben, *The New Yorker*

"Climate experts Ayana Elizabeth Johnson and Katharine K. Wilkinson gathered a chorus of 60 courageous women who are using their individual voices to create collective change around the climate crisis. The resulting anthology is an anthem for the world—at once a battle cry and unifying hymn, a mournful melody and a song of salvation that will ring true for generations to come."

—*Atmos*

"A fiery, hopeful manifesto on how to make sense of the staggering loss posed by climate change—and take justice-oriented action in spite of it."

—*Mashable*

"It offers a path forward, one where 'we,' in the truest meaning, grasp power and fix what is broken. The essays espouse how to restore our relationship with nature and each other. Together, they call for a decidedly humanist and inclusive approach welcoming everyone from wherever they are into the fight to protect what's left and repair what we can."

—*Gizmodo*

"A welcome anthology, in prose and verse, of women's writings on climate change . . . [Johnson and Wilkinson] write that the political and social constructs that oppress women are one and the same with those that are wreaking havoc on the global environment. . . . There's no such zero-sum game-playing here. The editors observe that women are well equipped to transcend ego and competition in order to create a politics of 'heart-centered, not just head-centered, leadership.' . . . A well-curated collection with many ideas for ways large and small to save the planet."

—*Kirkus Reviews*

"Hopeful and illuminating, *All We Can Save* is an anthology of essays by women at the forefront of the climate crisis. So often climate writing can make us feel doomed and anxious, but this collection is a comfort because of its honesty and courage. . . . It's a reminder that we can work with hope towards a better future."

—*BuzzFeed*

"The anthology is a tribute to the fearless activists, journalists, conservationists, and others who are bringing forth what Johnson and Wilkinson call a 'feminist climate renaissance,' rooted in the traditionally feminine qualities of compassion, connection, creativity, and collaboration."

—*Grist*

"Women of all backgrounds—artists, writers, scientists, policy makers, and others—are at the forefront of climate action, and with this exquisite anthology, marine biologist Dr. Ayana Elizabeth Johnson and editor-in-chief of Project Drawdown, Dr. Katharine Wilkinson, bring their voices together."

—*Literary Hub*

"The book is not only an entertaining and varied read, it's also a crucial catalyst for change—one that led to a new project that provides support and community for women climate leaders."

—*Smithsonian Magazine*

"A splendid offering of wisdom, warmth, and inspiration to reshape our vision of climate futures, *All We Can Save* is a skillfully curated collection of essays, poems, and illustrations that is decidedly feminine in its character and feminist in its approach. . . . A beautiful cross-section of society's most visionary voices on climate."

—*Yes! Magazine*

"This astounding and ambitious eco-anthology is filled with whip-smart essays, heart-wrenching poems, and stunning visual art from an all-female cast . . . those who've been left out of the climate debate for too long. . . . A powerful chorus of women armed with solutions for our changing climate."

—*Self*

"*All We Can Save* brings an empathetic perspective to a fraught subject . . . a community bound between two covers, and a gift to any who wishes to join in. Johnson and Wilkinson have set a high bar [with] this movement-forging book."

—Bloomberg Green

"These are the ideas, solutions, inspirations, and visionaries we've been waiting for."

—*Ms.*

ALL WE CAN SAVE

Truth, Courage, and Solutions for the Climate Crisis

EDITED BY

Ayana Elizabeth Johnson

&

Katharine K. Wilkinson

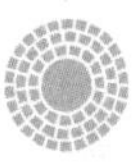

ONE WORLD • NEW YORK

2021 One World Trade Paperback Edition

Published in the United States by One World, an imprint of Random
House, a division of Penguin Random House LLC, New York.

One World and colophon are registered trademarks of Penguin
Random House LLC.

Credits appear on page 389.

Originally published in hardcover in the United States by One World,
an imprint of Random House, a division of Penguin Random
House LLC, in 2020.

Trade paperback ISBN: 978-0-593-23708-3
Ebook ISBN: 978-0-593-23707-6

This book was printed in the United States of America on
Rolland Enviro™ Book stock, which is manufactured using
FSC-certified 100% postconsumer fiber.

oneworldlit.com
randomhousebooks.com

3rd Printing

Book design by Simon M. Sullivan
Illustrations by Madeleine Jubilee Saito

For our mothers, our sisters, and Earth

My heart is moved by all I cannot save:
so much has been destroyed

I have to cast my lot with those
who age after age, perversely,

with no extraordinary power,
reconstitute the world.

—Adrienne Rich

Contents

4. RESHAPE

5. PERSIST

6. FEEL

7. NOURISH

8. RISE

Editors' Notes

Throughout the book, you will see marks in the margins to highlight some of the key statistics (*) and insights (•••••) we found poignant. We have also underlined some key terms and the names of women whose climate-related work is referenced. We have capitalized all races, following scholar Dr. Ibram X. Kendi and others.

To borrow words from Rachel Carson's *Silent Spring*, "I have not wished to burden the text with footnotes but I realize that many of my readers will wish to pursue some of the subjects discussed. I have therefore included a list of my principal sources . . . at the back of the book." We have done the same, and in addition to select sources shared here, you can find a full list of references online: www.allwecansave.earth/references.

What brought the two of us together initially was a shared commitment to building community around climate solutions. We believe the climate movement is only as strong as the relational web between us, which needs to be nurtured. We hope this book can be a spark for connecting, learning together, deepening our resolve, and joyously finding our places in the mighty "we" that's rising to secure a just and livable future. Resources for collective reading circles are available on our website: www.allwecansave.earth/circles.

As we midwifed *All We Can Save* into the world, it became clear that it needed to be more than a book. So, we are carrying forward the mission held in these pages through a nonprofit, The All We Can Save Project. For more information and to join us, visit: www.allwecansave.earth.

BEGIN

CAN YOU IMAGINE ALL OF US

TRUSTING EACH OTHER

WORKING TOGETHER

FOR OUR COMMON HOME?

MADELEINE JUBILEE SAITO

Begin

AYANA ELIZABETH JOHNSON AND
KATHARINE K. WILKINSON

Eunice Newton Foote rarely gets the credit she's due. In 1856 Foote theorized that changes in carbon dioxide in the atmosphere could affect the Earth's temperature. She was the first woman in climate science, but history overlooked her until just a few years ago.

Foote arrived at her breakthrough idea through experimentation. With an air pump, two glass cylinders, and four thermometers, she tested the impact of "carbonic acid gas" (the term for carbon dioxide in her day) against "common air." When placed in the sun, she found the cylinder with carbon dioxide trapped more heat and stayed hot longer.

From a simple experiment, she drew a profound conclusion: "An atmosphere of that gas would give to our earth a high temperature; and if as some suppose, at one period of its history the air had mixed with it a larger proportion than at present, an increased temperature . . . must have necessarily resulted." In other words, she connected the dots between carbon dioxide and planetary warming, and she did it more than 160 years ago.

Foote's paper, "Circumstances Affecting the Heat of Sun's Rays," was presented in August 1856 at a meeting of the American Association for the Advancement of Science and then published. For unknown reasons it was read aloud by Joseph Henry, secretary of the Smithsonian, rather than by Foote herself. That was three years before Irish physicist John Tyndall published his own more detailed work on heat-trapping gases—work typically credited as the foundation of climate science.

Did Tyndall know about Foote's research? It's unclear—though he did have a paper on color blindness in the same 1856 issue of *The American Journal of Science and Arts* as hers. In any case, we have to wonder if Eunice Newton Foote ever found herself remarking, as so many women have: "I literally just said that, dude."

Foote wasn't only a scientist. She was involved in the early movement for women's rights too. Her name appears on the list of signatories to the 1848 Seneca Falls "Declaration of Sentiments"—a manifesto created during the first women's rights convention in the United States—right below suffragist Elizabeth Cady Stanton. Foote's husband, Elisha, and abolitionist-philosopher Frederick Douglass also signed on, under "gentlemen." (Of note: John Tyndall opposed women's suffrage.)

Foote, it seems, was a climate feminist.

The same patriarchal power structure that oppresses and exploits girls, women, and nonbinary people (and constricts and contorts boys and men) also wreaks destruction on the natural world. Dominance, supremacy, violence, extraction, egotism, greed, ruthless competition—these hallmarks of patriarchy fuel the climate crisis just as surely as they do inequality, colluding with racism along the way. Patriarchy silences, breeds contempt, fuels destructive capitalism, and plays a zero-sum game. Its harms are chronic, cumulative, and fundamentally planetary.

And these structures are being actively upended. The People's Climate March and the Women's March. School strikes for climate and the #MeToo movement. Rebellions against extinction and declarations that time's up. More than concurrent, these phenomena are connected by the systems they seek to transform and the values that guide them.

The climate crisis is not gender neutral. Climate change is a powerful "threat multiplier," making existing vulnerabilities and injustices worse. Especially under conditions of poverty, women and girls face greater risk of displacement or death from extreme weather disasters. Early marriage and sex work—sometimes last-resort survival strategies—have been tied to droughts and floods. There is growing proof of the link between climate change and gender-based violence, including sexual assault, domestic abuse, and forced prostitution. Tasks core to survival, such as collecting water and wood or growing food, fall on female shoulders in many cultures. These are already challenging and time-consuming activities; climate change can deepen the burden, and with it struggles for health, education, and financial security.

The list of harmful impacts caused by our rapidly changing climate goes long and it goes wide, especially for girls and women of color, those in the Global South, and those who are rural or Indigenous. In very real ways, the climate crisis thwarts the rights and opportunities of women and girls, as well as nonbinary people. These realities make gender-responsive strategies for climate resilience and adaptation critical. And they mean that bold climate action is critical to our aspirations for gender equality and justice.

However, the story does not, and must not, end with the label "victim." When you're close to the problem, you're necessarily close to the solutions.

All around the world, women and girls are making enormous contributions to climate action: conducting research, cultivating solutions, creating campaign strategy, curating art exhibitions, crafting policy, composing literary works, charging forth in collective action, and more. Look around and you will see on the rise climate leadership that is more characteristically *feminine* and more faithfully *feminist*, rooted in compassion, connection, creativity, and collaboration. There is a renaissance blooming in the climate movement, and it has a few important characteristics.

First, there is a clear focus on making change rather than being in charge. We see women and girls moving beyond ego, competition, and control, which are rampant in the climate space (as elsewhere) and impede good work. We see joyful following where wise leadership appears, joining instead of duplicating, giving one another credit, sharing resources, passing the mic, and celebrating one another's successes. It is shine theory in practice.

Second, there is a commitment to responding to the climate crisis in ways that *heal* systemic injustices rather than deepen them. We see women and girls centering justice, inclusion, and frontline communities, recognizing that we can address near-term needs and long-term aims at the same time, and more effectively. Equity is not secondary to survival, as some suggest; it is survival.

Third, there is an appreciation for heart-centered, not just head-centered, leadership. We see women and girls bringing their whole selves to this movement—fear, grief, fiery courage, wracking uncertainty, all of it—and doing the inner work that often precedes effecting change. The climate crisis has inescapable psychological and spiritual dimensions. What's so powerful about integrating head and heart: It's where scientific rigor and moral clarity, analysis and empathy, strategy and imagination meet. It is what allows us to sustain bold aspirations and insist upon the action that's necessary rather than what's expedient or "practical."

Fourth, and perhaps most important, there is a recognition that building community is a requisite foundation for building a better world. We see women and girls engaging in deeply relational, collaborative, and supportive ways—taking the necessary time, making the necessary space, investing in the weft and weave between us. It is clear that we are in this together, that our fates are intertwined. And in many ways, success requires building the largest, strongest team possible.

While women and girls are undeniably vital voices and agents of change for this planet, we are too often missing or even barred from the proverbial table. Women remain underrepresented in government, business, engineering, and finance; in executive leadership of environmental organizations, United Nations climate negotiations, and media coverage of the crisis; and in the legal systems that create and uphold change. Girls and women leading on climate receive insufficient financial backing and too little credit. Again, unsurprisingly, this marginalization is especially true for women of the Global South, rural women, Indigenous women, and women of color. The dominant public voices and empowered "deciders" on the climate crisis continue to be White men.

More than a problem of bias, suppressing the climate leadership and participation of women and girls—half the world's brainpower and change-making might—sets us up for failure. Research shows that women have an edge over men when it comes to the planet: caring about the environment and climate change and acting on that care; aversion to taking on outsized risk or imposing it on others (something data indicates White men are particularly inclined to do). This edge carries into politics and policy making. Female legislators more strongly support environmental laws—and stricter laws at that. When parliaments have greater representation of women, they are more likely to ratify environmental treaties. When women participate equally with men, climate policy interventions are more effective. At a national level, higher political and social status for women correlates with lower carbon emissions and greater creation of protected land areas. It's not about *only* women but about making sure women are included and leading at all levels.

To address our climate emergency, we must rapidly, radically reshape society. We need every solution and every solver. As the saying goes, to change everything, we need everyone. What this moment calls for is a mosaic of voices—the full spectrum of ideas and insights for how we can turn things around.

The climate crisis is a leadership crisis. For far too long, far too many leaders have been focused on profit, power, and prestige; and many of those committed to change have been ineffective. The climate crisis is the result of social, political, and economic systems that are wildly skewed to benefit those who already have so much. It is the result of unfettered economic growth, extractive capitalism, and the concentration of wealth and power in the hands of a few, who have known plenty but cared too little and continue to block efforts for change. Humanity simply cannot survive the status quo, and neither can many other species and ecosystems.

To transform society this decade—the clear task science has set before us—we need transformational leadership. We need feminine and feminist climate leadership, which is wide open to people of any gender. This is where possibility lives—possibility that we can turn away from the brink and move toward a life-giving future for all.

What might Eunice Newton Foote have achieved if she had had John Tyndall's access to training and resources, the same platform and power? We can only imagine. And while we cannot rewrite that chapter, we can hear, heed, and support climate feminists today.

This book imparts the minds and hearts of women leading the way on climate and using their diverse gifts to, as Adrienne Rich writes, "reconstitute the world." They are a small subset of a mighty community showing up in this moment, gathering the pieces of a broken world, doing the work of mending.

The climate crisis is inescapably international, but this collection intentionally focuses on the United States as the country most responsible for causing it and with so very much work to do. For too long Americans have tended to compartmentalize climate change as something that affects (poor, Black, and Brown) people far away, something that is too bad but not our problem. Well, that bubble has burst: hurricanes in the Gulf and Caribbean and along the Eastern Seaboard; fires on the West Coast; floods in the Midwest; droughts here, deluges there, and heat waves everywhere; diseases spreading; insects and songbirds disappearing; sea level ever rising; erratic weather making it harder to grow food. Our visionaries, many of them women and people of color, have been not only warning us but illuminating paths forward.

This book initially had a dual aim: to shine a light back on them, uplifting the expertise and voices of dozens of diverse women leading on climate in the United States—scientists, journalists, farmers, lawyers, teachers, activists, innovators, wonks, and designers, across generations, geographies, and races—and to advance a more representative, nuanced, solution-oriented public conversation on the climate crisis. As we put it together, intermixing essays with poetry and art, as all these voices became a chorus, this book also became a balm and a guide for the immense emotional complexity of knowing and holding what has been done to the world, while bolstering our resolve never to give up on one another or our collective future.

The writings in this collection are united by their willingness to grapple with big questions: How did we get into this mess? What is at stake? How can we make sense of this crisis, psychologically and spiritually? What solutions exist from individual to national levels? How can we ensure justice is embedded in transformation? What civic and cultural shifts can help turn things around? What do we need to do? Whom do we need to become? What might possible futures look like? How can we reach them together? The necessary exploration, ideas, and solutions, of course, go beyond what these pages could contain. The answers shared here are expansive but not exhaustive.

The book unfurls in eight parts, their essences captured here in light brushstrokes.

Root

A call, a welcoming, a place to ground
The foundation of Indigenous wisdom

And the wisdom of Earth's living systems
Interconnection, emergence, justice, regeneration

Advocate

Strategy, participation, public good
Plying tools of legislation and litigation
How we hold the powerful to account
And (re)write the rules with all people in mind

Reframe

Language and story, creativity and culture
Our means of making sense
To tell the truth—expand, flip, and rekindle it
Imagining, evolving, holding on to our humanity

Reshape

Problems embedded in the contours
Of cities, transport, infrastructure, capitalism
Coastlines and landscapes where human-nature meet
Much to reconsider, rend, invert, remake

Persist

Damn if this work isn't hard, our task towering
The fire of activism—on the front lines, in the belly
Standing for justice, for health, for the sacred
We don't have to do this alone

Feel

Awake, aware, attuned
Hearts break, souls shake with anxiety
Can't skip this: struggling, mourning, raging, healing
A ferocious love for the planet we call home

Nourish

Soil, food, water, sky—inseparable
The foundations of our aliveness
Collaborating with and supporting nature
Microbes, farmers, photosynthesis

Rise

Generations—growing, giving, gathering
Nurture community and transformation
For a future that holds us, all of us
This is the work of our lifetimes

This book is about a spectrum of work that needs doing and a collective effort to make our best contributions; it's not about heroes. So whether you are a veteran of the climate movement, a keen onlooker from the sidelines, or someone joining this conversation for the first time, we hope you will find yourself in these pages. We have peeled away jargon, included foundational information, and created simplicity without forfeiting complexity, because this book is for *everyone* concerned about our shared future. For everyone seeking fresh perspectives and bold ideas. For those who value diverse voices on the pressing issues of our day. For those already involved and those still discovering what role they might play.

This book refuses to dodge how bad things are, yet keeps a forward gaze. Because movements don't alter history by saying: *What if we don't succeed? . . . Some things will never change. . . . The odds seem really long. . . . Maybe we'll never get the right to vote, to marry, to be free. . . .* Without knowing the outcome, we have to try anyway; without a single guarantee, we must show up. So we focus on how to understand where we are and where we go from here. As the subtitle lifts up, we must summon *truth, courage, and solutions.* This trifecta can move us forward, through the aching uncertainty.

In the words of adrienne maree brown, "What we pay attention to grows, so I'm thinking about how we grow what we are all imagining and creating into something large enough and solid enough that it becomes a tipping point." And what we want to tip toward is community, care, repair, and renewal. We want to tip toward life. While it is too late to save everything—some ecological damage is irreparable, some species are already gone, ice has already melted, lives have already been lost—it is far too soon to give up on the rest. So let these women, these visionaries, lead you on a path toward all we can save.

1. ROOT

MADELEINE JUBILEE SAITO

Calling In

Xiye Bastida

I was born and raised in a small town called San Pedro Tultepec, outside Toluca, a city in the basin that neighbors Mexico City. Between 2011 and 2013, my town experienced the worst drought in Mexico in seventy years. Then in 2015 we were hit by heavy rainfall that resulted in flooding, affecting worst the poor communities by the Lerma River. This was the first time I saw the effects of our climate crisis and how unjust they are.

When my family and I later moved to New York City, I saw what Superstorm Sandy had done to Long Island and the seashore, and I thought: *It's all the same problem.*

The climate crisis can feel really complicated and overwhelming at times. Carbon emissions keep increasing. Pollution of the air, water, and soil makes us sick. Some people just shut it all out. Others refuse to accept the scientific consensus around the subject. And unsurprisingly, most people have not read the United Nations Intergovernmental Panel on Climate Change (IPCC) reports and don't understand how the greenhouse effect works. Some people feel like they have to know the science inside out before they can talk about it or do something about it. But here's what I have learned: *You don't have to know the details of the science to be part of the solution.* And if you wait until you know everything, it will be too late for you to do anything. That's why we, the youth who are leading on climate, are calling this an *emergency*.

Many people are now giving all the credit to young people when looking for key players in the climate movement. But as youth, we know that we didn't start the movement at all. And we also know that environmental consciousness didn't start in 1962 with Rachel Carson's *Silent Spring* or when Earth Day began in 1970. We are just bringing the element of urgency because we realize that we are running out of time. In 2018 the IPCC gave us a deadline: We have ten years to nearly halve global carbon emissions to stay below 1.5 degrees Celsius of warming from preindustrial levels. *

Youth have come to understand that we have to be the communicators of science, facts, policy, solutions, and hope through language that reaches the general public. We have to use every tool at our disposal, from traditional media to memes, to tell the world what we know and why it matters to us.

Older generations were able to enjoy and do whatever they wanted without worrying about climate refugees and extreme climate disasters. But now the world around us is deteriorating beyond a point of natural recovery, and we are not going to be able to inhabit a world where nature is healthy and in balance. It's as unfair to ecosystems as it is to humans. That's why we are rising up.

Our generation holds a kind of consciousness that is not based on monetary gain or on new ways of profiting from lands, forests, rivers, seas, and people. We are pushing for a complete breakthrough of sensible and wise solutions that ensure the continuation of ecosystems and peaceful societies. For this, human civilization needs to make drastic changes in its value system; it needs to mature.

As a descendant of the Otomi-Toltec people, I feel my elders have the kind of guidelines and principles that humanity needs in these critical times. I was raised with the philosophy of my ancestors: that you take care of the Earth because she takes care of you. Indigenous peoples have been doing that for thousands of years; it is their culture and their way of life. It is clear within the youth movement that being a climate justice activist or an environmentalist working for the regeneration of Earth will require continuous dedication. It can't be a hobby; it has to be a shift in culture and mindset.

We don't want people to have to experience climate disasters to realize that this is a crisis. We want enough youth to see the power of collaborating across sectors and generations so that we change the conversation toward solutions. We want all communities to embody the morality and wisdom needed to adapt to a post–fossil fuel era. If more people become aware of what's going on, they will make radical decisions and take radical action.

We need to have a whole cultural shift, where it becomes our culture to take care of the Earth, and in order to make this shift, we need storytelling about how the Earth takes care of us and how we can take care of her.

Part of this will be learning to work by the principle of collaboration rather than competition. It's so different from what we've been taught—to be individualistic and to strive for personal success. In the movement I'm helping to build, we make all our decisions through consensus; there's no hierarchy. We appreciate one another's time, energy, and drive to change the world.

When I was younger, I envisioned companies as being surrounded by impenetrable gray walls. I thought that I couldn't change corporations because they were inaccessible, and I thought they didn't *want* to change. But now youth understand that companies are made up of *people*. In 2019, for the first massive, global youth climate strike, we saw some corporations close their stores, shut down their websites, and publicly support the strike. We watched people working at those companies speak up; for example, thousands of Amazon workers walked out for climate justice. We realized that we could change the narrative of corporate greed and incite businesses to divest from fossil fuels while striving for coexistence with Earth. Our approach as young consumers is to say, "We're gonna support you more if your business makes changes for the better."

It's true that the climate-strike movement is organized and led by youth, but we need to work intergenerationally if we want to impact every sector and every industry. To me and a lot of other young people, it feels like we're rooted in awareness while the adults around us live in obliviousness. This is where "Okay, Boomer" came from, a phrase designed to describe the intergenerational disconnect of the movement. But that's not right. First, the world is *not okay*, Boomer, so we will shake those who need to be shaken out of their comfort zone. People in developed countries and big cities are too comfortable, and nothing changes when we stay in a state of unbotheredness. We need to be uncomfortable about the system to change it, and the youth are bringing the climate crisis to attention precisely to channel action out of uneasiness. Second, we cannot let phrases like "Okay, Boomer" divide us. The fossil fuel industry wants us to be divided in order to slow down the push for climate justice. But we refuse to let attempts at division affect our purpose.

Our movement for climate justice isn't just led by young people: It's led by young people from all over the world, of all races and ethnicities. The Western media created a narrative that the youth movement started in Europe with Greta Thunberg; thus, people often see it as a White movement. But youth of color, who are disproportionately affected by the climate crisis, are also at the forefront of the movement. There are incredible environmental justice initiatives and organizations all over the world, big and small, where people are on the front lines of "natural disaster zones," vulnerable to extreme weather. They are all fighting for their rights to be better prepared but also to have more resilient ecosystems through passing bills that restrict or suspend mining, logging, damming, fracking, and other extractive activities.

The predominantly White, mainstream environmental movement is starting to acknowledge all the Indigenous, Black, and Brown communities on the front lines. These efforts are merging in a constructive way that will make the movement stronger.

We have helped propel the narrative transition from a call for climate action to the necessity of climate justice. And if you want to join us, here are my ten tips for being a climate justice activist:

1. Don't start from scratch. There are hundreds of existing initiatives that you can join.
2. Maintain good communication with your peers and the adult organizations you partner with.
3. Take good care of yourself and others.
4. Make your activism intersectional; include all stakeholders in your decision making, and don't tokenize.
5. Don't do things the patriarchal way, the racist way, the exhausting way, or any way that excludes marginalized voices just to be "efficient" and do things fast.
6. At events you hold, invite Indigenous peoples to do land acknowledgments, and remember that Indigenous knowledge is the foundation for addressing the climate crisis.
7. Always convey that individual and structural change are *both* indisputably necessary.
8. Meet people where they're at. Not everyone knows the climate crisis back and forth. Explain it and present solutions.
9. Use accessible language. Not everyone knows about ppm (parts per million) or the IPCC (Intergovernmental Panel on Climate Change).

10. Talk about greenwashing, environmental racism, green gentrification (or as I call it, *green*trification), and what a just transition means.

We as youth agree on the fact that addressing the climate crisis is a survival issue for everyone. Climate change doesn't discriminate, but the ability to respond to climate disasters does. Wildfires in California, for instance, are not avoiding fancy homes. Snowstorms, hailstorms, and tornadoes are not discriminating either. But fully recovering from these increasingly unnatural disasters is very difficult, almost impossible, for communities of color, Indigenous reservations, small rural towns, and traditional fishing communities.

Our local representatives should pay attention to every opportunity to reverse the policy and economic conditions that put our communities and ecosystems in peril. That's why we, as youth, lobby at the local level. You can do it too. Go to your state capitol and tell your representatives: "The climate crisis is an issue that we care about, and you need to represent that in your policy." When representatives see that our main concern is the climate crisis, they may feel obliged to act because they have kids too, and at the end of the day, they are public servants for the people's needs.

It's time to change our mindset toward implementing solutions. A vibrant, fair, and regenerative future is possible—not when thousands of people do climate justice activism perfectly but when millions of people do the best they can.

Reciprocity

Janine Benyus

Midway through my forestry degree, I found myself pointing a can of spray paint at the smooth bole of an ironwood tree. I was to mark it as part of a "release cut" in our experimental forest in New Jersey. The orange slash would tell loggers to fell, poison, or girdle anything that might compete with our sawtimber crop. We were taught that thinning would help the oaks and walnuts, freeing them to get more water, light, and nutrients. For many in our class, "opening up" a stand of trees was their favorite part. For me, it was an excruciating, empty choice.

I kept envisioning the historic forest right next to ours that hadn't been cut for two hundred years. I had seen overstory giants grouped in twos and threes and fours, a middle layer of hardwoods and conifers, and, at my feet, trilliums, fiddleheads, and rufous-sided towhees bursting from the duff. Nobody had released these trees from competition, yet all appeared well.

"The old forest is not nearly as open or regimented as this," I told my professor, "but it looks healthier. Do you think the trees might be grouped together for a reason? Do you think they might be benefiting one another in some way?"

He shook his head no, a bit alarmed. "Don't be so Clementsian," he said. "You'll never get into grad school." The reference was to Frederic Edward Clements, an ecologist from the early 1900s who had won and then lost the greatest debate in ecological history. Being compared to Clements was a well-known admonition, a sure sign of naïveté.

It was 1977, and ecologists were three decades deep into a paradigm shift that affected our experiments, our narratives about the wild, and, most powerfully, our maxims for managing forestlands, ranches, and farms. The precept that trees needed to be released from the struggle of competition was the fruit of a debate between Frederic Clements and his contemporary Henry Gleason. What they both endeavored to describe, in very different ways, was what constitutes a community of vegetation, what determines how plants grow together and why.

When Clements studied bayous, chaparrals, hardwood forests, and prairies, he saw distinct communities of plants reacting not just to soils and climate but also to one another. He proposed that plants were cooperators as well as competitors, facilitating one another in beneficial ways. Canopy trees "nursed" the saplings beneath their branches, creating more sheltered, nutritious conditions in a plant-helping-plant process later dubbed *facilitation*. They shaded seedlings from the drying sun, blocked the winds, and fertilized the soil with their leaves. As time passed, one community of plants prepared the way for another; annual plants built the soil for perennial shrubs, and those shrubs nourished saplings that grew into forests. Everywhere Clements looked, he saw communities so tightly interwoven, he called them organismic.

Gleason had a different take. What Clements called communities was simply happenstance, random individuals dispersed by chance and arranged according to how they adapted to water, light, and soil. There was no mutual aid; plants were merely competing for a spot in the struggle. The notion that there might be a connected, interdependent community to be studied as a whole was an illusion; examining the parts would do.

For the first half of the twentieth century, Clements's view prevailed—the ecological literature was full of studies on facilitation. Gleason's work was virtually forgotten until 1947, when a small group of researchers resurrected his individualist views and pitted them against Clements's holism. Ecologists in the throes of "physics envy" liked the fact that Gleason's view of plants allowed individuals to be studied with neat statistical precision, like atoms.

Within twelve short years, Clements's cooperative-community theory all but disappeared from the scientific literature, and the majority of ecologists rejected the idea of positive interactions as a driver of community assembly. Graduate students steered clear of the "unscientific aura" of Clements's ideas and instead focused their research on antagonistic interactions such as competition and predation. Given the times, we shouldn't be surprised. Clements's fall from grace coincided, to the year, with the release of the Truman Doctrine and the onset of the Cold War. For decades, communism was a third rail best avoided, even when talking about plants.

But here is what I love about the scientific method. Though culture seeps into science and sometimes holds its finger on the scale, it cannot stop the restless search for measurable truth. Un-American or not, the math has to work. When fifty years of wall-to-wall research into competition proved inconclusive, researchers went back to the field to find out what else was at play.

The same year I was pardoning an ironwood, ecologist Ray Callaway was in the foothills of the Sierra Nevada, rescuing blue oaks from bad practice. The prevailing wisdom, a gift from Gleason, was that the oaks dotting California's rangelands should be cut to release grasses from competition. Much to Callaway's dismay, thousands of acres of blue oaks were being bucked up for firewood.

The fact that grasses had thrived with blue oaks for eons nagged at him. How harmful could oaks be? For two and a half years, he measured the interaction between the oaks and the grasslands—his pans and buckets catching leaves, twigs, branches, and nutrient-laced rainfall dripping from oak canopies. His thesis showed that the nutrient totals under oaks were actually five to sixty times greater than in open grasslands. Those spreading trees, so artfully arranged in the California landscape, are nutrient pumps that lift minerals from the deep and scatter them in an annual leaf drop. Penetrating taproots loosen the dense soil, increasing water storage beneath the boughs and welcoming a profusion of plants. Today experts recommend protecting, not cutting, these "islands of fertility."

Callaway has gone on to become a leading light in facilitation research. His book *Positive Interactions and Interdependence in Plant Communities* reviews more than a thousand studies of how plants chaperone and enhance their neighbors' growth, survival, and reproduction. To read these strategies is to discover a manual for how life evolved on a challenging planet and how natural communities heal and overcome adversity—essential reading for a climate-changed world.

Knowing which plants are the chaperones in botanical communities will be important as droughts deepen in the coming years. For example, how and why do Amazonian rainforests create clouds even in the dry season? It turns out that 10 percent of the Amazon's annual rainfall is absorbed by certain trees and shrubs. Their shallow roots absorb the rainfall, and their deep taproots push it down deep into the soil bank. When the rainless months come, water rises up the taproots and gets distributed to the whole of the forest. Many species of plants throughout the world perform this hydraulic "lift," watering a multitude of plants under the forest canopy.

The more stressful the environment, the more likely you are to see plants working together to ensure mutual survival. On Chilean peaks, studies of mounded plants huddling together against harmful ultraviolet rays and cold, drying winds reveal complex interactions of support. A single six-foot-wide yareta, or cushion plant, can be thousands of years old and harbor dozens of different flowering species in its mound, tucked like colorful pins in a bright green cushion.

Downslope, tear-shaped "tree islands" show how facilitation can create the conditions conducive to community. If a tree can tough it out and get established on a rockfall, it creates a lee where winds calm and snows drift to water sheltered seedlings. Decaying leaves and needles create an organic sponge that collects and then releases moisture in the dog days of summer. Birds roost and mammals hide in the growing island, importing nutrients and seeds in their excrement. Over centuries, these tree islands migrate, the windward sentries succumbing while the leeward ranks march backward. Studies have shown that tree islands act as mobile soil builders, roaming back and forth, painting fertility across mountainsides.

Whether it comes in the form of shading, shielding, nourishing, or defending, facilitation allows plants to expand their niches, to thrive where they would normally wither. Landscapers, farmers, and foresters may want to mimic these moves by planting for partnership, including wind blockers, soil holders, water lifters, and nutrient boosters in their mixtures. As plants deal with shifting growing zones, a facilitation partner could make all the difference.

I admit it's counterintuitive to imagine plants increasing facilitation in the face of scarcity when our competition bias and our economic theories suggest otherwise. For years, careful experimenters tried to explain this as an anomaly, missing the beneficence in their search for the struggle. Now we know that it's not just one plant helping another; mutualisms—complex exchanges of goodness—are playing out above- and belowground in extraordinary ways.

While Callaway was measuring oaks in California, Suzanne Simard was a professional forester, wincing through British Columbia's mass clear-cuts. The management protocol of removing paper birch trees that grew in association with Douglas fir seemed strange to her—they had been companions for eons. Might they be helping each other in some way?

In a brilliant study, she exposed growing seedlings to two types of carbon dioxide—radioactive carbon-14 for birch and stable carbon-13 for Douglas fir. The seedlings would absorb the carbon dioxide and transform it into sugars. She followed the carbon to see if any would be exchanged. The first results came in an hour's time. She describes a sense of wonder bordering on euphoria when the Geiger counter popped and clicked—carbon-14 from the birch had traveled to the Douglas fir, while carbon-13 from the fir made its way to the birch.

How? Next time you're in a forest, dig into the duff, and you're bound to find white, cobwebby threads attached to roots. These are the underground parts of special fungi that deliver phosphorus to trees in return for carbon. Textbooks once described the exchange as exclusive, one tree to one fungus, until the data begged to differ. Simard's work was among the first to prove that fungi branch out from the roots of one tree to connect dozens of trees and shrubs and herbs—not only relatives but entirely different species. This "wood-wide web" is an underground Internet through which water, carbon, nitrogen, phosphorus, and even defense compounds are exchanged. When a pest troubles one tree, its alarm chemicals travel via fungi to the other members of the network, giving them time to beef up their defenses. Thanks to researchers like Simard, foresters are now encouraged to leave birch and large hub trees in the forest to give seedlings a fast connection to the network.

Discoveries about the connected nature of mutualists have vast implications for forestry, conservation, and agriculture in a warming world. Although 80 percent of all land plants have roots that grow in association with mycorrhizae fungi, it's rare to find thriving mycorrhizal networks in agricultural fields. Plowing disturbs the cobwebby network, and the year-on-year addition of artificial nitrogen and phosphorus fertilizers tell bacterial and fungal helpers that they are no longer needed—not needed for water transport or pest defense, not needed to absorb the micronutrients our bodies long for. It's time to bring the wood-wide web to farmlands.

When communities of vegetation breathe in carbon dioxide, turn it into sugars, and feed it to microbial networks, they can sequester carbon deep in soils for centuries. But to do that, the communities need to be healthy, diverse, and amply partnered. If we're to encourage wild and working landscapes to recoup the 50 to 70 percent of soil carbon that has been lost to the atmosphere, we'll want to pause before plowing a field, opening a bag of fertilizer, or marking a sapling for removal. We wouldn't want to interrupt a vital conversation. *

If humans are to help reverse global warming, we will need to step into the flow of the carbon cycle in new ways, stopping our excessive exhale of carbon dioxide and encouraging the winded ecosystems of the planet to take a good long inhale as they heal. It will mean learning to help the helpers, those microbes, plants, and animals that do the daily alchemy of turning carbon into life. This mutualistic role, this practice of reciprocity, will require a more nuanced understanding of how ecosystems actually work. The good news is that we're finally developing a feeling for the organismic, after years of wandering in the every-plant-for-itself paradigm.

One of the fallouts of our fifty-year focus on competition is that we came to view all organisms as consumers and competitors first, including ourselves. Now we're decades into a different understanding. By recognizing, at last, the ubiquity of sharing and chaperoning, by acknowledging the fact that communal traits are quite natural, we get to see ourselves anew. We can return to our role as nurturers, each a helper among helpers in this planetary story of collaborative healing.

The Big Picture

ELLEN BASS

I try to look at the big picture.
The sun, ardent tongue
licking us like a mother besotted

with her new cub, will wear itself out.
Everything is transitory.
Think of the meteor

that annihilated the dinosaurs.
And before that, the volcanoes
of the Permian period—all those burnt ferns

and reptiles, sharks and bony fish—
that was extinction on a scale
that makes our losses look like a bad day at the slots.

And perhaps we're slated to ascend
to some kind of intelligence
that doesn't need bodies, or clean water, or even air.

But I can't shake my longing
for the last six hundred
Iberian lynx with their tufted ears,

Brazilian guitarfish, the 4
percent of them still cruising
the seafloor, eyes staring straight up.

And all the newborn marsupials—
red kangaroos, joeys the size of honeybees—
steelhead trout, river dolphins,

so many species of frogs
breathing through their damp
permeable membranes.

Today on the bus, a woman
in a sweater the exact shade of cardinals,
and her cardinal-colored bra strap, exposed

on her pale shoulder, makes me ache
for those bright flashes in the snow.
And polar bears, the cream and amber

of their fur, the long, hollow
hairs through which sun slips,
swallowed into their dark skin. When I get home,

my son has a headache and, though he's
almost grown, asks me to sing him a song.
We lie together on the lumpy couch

and I warble out the old show tunes, "Night and Day" . . .
"They Can't Take That Away from Me" . . . A cheap
silver chain shimmers across his throat

rising and falling with his pulse. There never was
anything else. Only these excruciatingly
insignificant creatures we love.

Indigenous Prophecy and Mother Earth

SHERRI MITCHELL
WEH'NA HA'MU KWASSET, PENAWAHPSKEK NATION

I used to keep a cartoon image on the wall in my office. The image was of two scientists with clipboards standing over a maze. Inside the maze was a mouse who had successfully reached the end and found the cheese. The caption read: "He found the cheese again. He loves it in there."

I kept that cartoon to remind me that this same type of biased science was used throughout history to paint a distorted view of Indigenous peoples and our ways of being and knowing. Racially biased science has been used to dehumanize and diminish Indigenous peoples in the eyes of the larger society from the early days of scientific exploration.

Though research is always influenced by the biases and beliefs of the researcher, this is especially true when there are significant cultural differences between the observer and those being observed. Assuming scientific neutrality when it comes to the study of cultures that are vastly different from one's own is naive at best. And when those studies are initiated for the purpose of proclaiming one culture superior to another, the hope for neutrality becomes even more distant. Lamentably, history has shown that it was not bias alone that influenced scientific research on culture and race, but something more directed and nefarious. On December 27, 1899, British major Ronald Ross informed members of the Liverpool Chamber of Commerce that "the Great Powers, tired of self-development, were endeavoring to extend their possessions and civilization all over the world," and that he believed imperial success would largely depend upon scientific success. A hundred years later, we learn that Major Ross's declaration was a signal for what would come. In her stunning book *Decolonizing Methodologies*, Indigenous scholar Linda Tuhiwai Smith writes, "The ways in which scientific research is implicated in the worst excesses of colonialism remains a powerful remembered history. . . . This collective memory of imperialism has been perpetuated through the ways in which knowledge about Indigenous peoples was collected, classified and then represented."

Racist ideologies are not only responsible for the aberrant social thinking that led to the exploitation of Indigenous peoples. They are also responsible for the creation of a racially exclusive framework that has bolstered colonial scholarship and relegated Indigenous knowledge to obscurity. The purposeful degrading of Indigenous knowledge by mainstream governments, academics, and scientists has led to distorted ideas about our intellect and created countless stereotypical myths about our ways of knowing and being. Because Indigenous peoples didn't share European ideas about land ownership, we were considered primitive. Because we had no desire to place the sources of our survival ("natural resources") into the stream of commerce, we were viewed as ignorant. And because our value system was based on relationships and not currency, we were believed to lack the capacity to live "civilized" lives. Ironically, the Indigenous ways of knowing and being that European colonists saw as primitive and uncivilized are now being actively sought out to save our environment and humankind from the brink of extinction.

Indigenous knowledge is based on millennia-long study of the complex relationships that exist among all systems within creation. It encompasses a broad array of scientific disciplines: ethnobotany, climatology, ecology, biology, archaeology, psychology, sociology, ethnomathematics, and religion. The keepers of Indigenous knowledge carry thousands of years of data on things such as medicinal plant properties, biodiversity, migration patterns, climate changes, astronomical events, and quantum physics. They carry the stories of countless epochs of human history, going all the way back to the beginning of human life on Mother Earth. And they provide insights that help fill the gap between our physical and subjective experiences, enabling us to understand how our internal consciousness impacts the ways that we view and experience the world around us.

In recent years, many scientists have realized that they are just now "discovering" what Indigenous peoples have long known. For instance, archaeologists and environmental scientists have found Indigenous marine-management sites that predate European settlement; they had believed marine management to have come only after European influence, despite information to the contrary that was provided by Indigenous peoples. Scientists are now gaining valuable information regarding changes in the local ecosystem from the Indigenous peoples in some coastal areas, who are helping them to develop updated management plans. In addition, Indigenous oral traditions have corrected inaccurate or misleading accounts of major events in U.S. history, providing complete and accurate depictions of interactions between the U.S. Cavalry and various Indigenous nations, including detailed drawings of battle sites that chronicle everything from wound locations to the colors of the uniforms.

The challenge for science has always been to see beyond the confines of its inherent bias and to overcome hierarchical, reductionist, and compartmentalized thinking to see the holistic patterns that are present throughout creation. Seeing the world through an Indigenous lens requires one to take a world-centered view that recognizes the relationships that exist among all living systems and the many ways that these systems are constantly moving toward harmony and balance. Unfortunately, a great deal of critical Indigenous knowledge has remained outside the carefully ordered categorization of Western thought, making its holistic concepts difficult to comprehend for those who have been trained to see the world in fractured pieces. It is this fractured view that has been central to the fracturing of our societies and environment.

Indigenous kinship systems, with their inclusion of beings from the natural world, have been viewed as little more than magical thinking by mainstream science. A. S. Thomson, a doctor in the British army who authored volumes of scientific publications, wrote the following about Indigenous peoples: "The faculty of imagination is not strongly developed among them, although they permitted it to run wild in believing absurd superstitions." He was not alone in his beliefs. The extended kinship networks of Indigenous peoples were viewed as a superstitious absurdity by colonial scientists for centuries and were often cited as justification for the paternalistic practices of colonial governments. However, in 2015 a significant shift occurred. In September of that year, scientists from eleven institutions published the first draft of their Open Tree of Life Project in *Proceedings of the National Academy of Sciences.* This phylogenetic map, funded by the National Science Foundation, shows the living connection among 2.3 million species. This was the first time that science acknowledged that our kinship teachings were more than fanciful fiction. Today interest in Indigenous knowledge is growing within the scientific community, now that there is increasing awareness that it accurately describes a number of unfolding scientific theories.

Indigenous knowledge recognizes the individuality of elements in the natural world and how they relate to a larger whole using traditional family kinship models as their scaffold. It does this without stripping away the individual value or attempting to force what is being seen into a larger body of generalized laws or theories. It simply recognizes the familial relationship and acknowledges that all life is both sovereign and interdependent, and that each element within creation (including humans) has the right and the responsibility to respectfully coexist as coequals within the larger system of life. If we hope to truly align Indigenous knowledge with Western thought in order to address the crises of our time, mainstream thinkers will have to be courageous enough to challenge the barriers that have prevented Indigenous knowledge from coming forward previously and begin to expand their sight to a more holistic vision. And since all scientific discovery stands on the shoulders of those who have come before, the field of science itself will have to reconcile its shameful past and revisit the reprehensible displays of racially driven confirmation bias that form its foundations.

Today we face multiple systems on the verge of collapse, including vital ecosystems that are necessary to sustain life. Experts from the United Nations, the scientific community, governmental agencies, and environmental organizations have all pointed to Indigenous peoples as being key to addressing climate change. Those who once claimed that Indigenous issues were fringe or ephemeral are now recognizing that protecting Indigenous rights is an ecological necessity.

Though Indigenous peoples comprise only about 5 percent of the * global population, our lands hold approximately 80 percent of the world's biodiversity and an estimated 40 to 50 percent of the remaining protected places in the world. We also create the least greenhouse gases and have the largest carbon stores on the planet within our territories. This makes the protection of Indigenous rights, along with our lands and ways of life, critical to the well-being of the planet, and it makes us natural partners in the race to address climate change. But we must be discerning in those partnerships. There are many who appear to be seeking Indigenous partnerships merely to overcome the obstacles that Indigenous rights present to ongoing development.

We must also recognize that climate change is only one symptom of a larger problem. Human beings have fallen out of alignment with life. Their beliefs and ways of being have shifted dramatically from those of their ancestors, taking them further and further away from the sources of their survival. As a result, people have forgotten how to live in relationship with the rest of creation. They have lost their respect for their elders in the natural world, such as the trees, waters, soils, and millions of other species that thrived on Mother Earth long before human beings arrived. Therefore, the greatest contribution that Indigenous peoples may be able to make at this time is to continue providing the world with living models of sustainability that are rooted in ancient wisdom and that inform us how to live in balance with all of our relations on Mother Earth. This will require non-Indigenous people to stand with us and ensure that our lands, waters, and ways of life are not further eroded by government and industrial intrusion.

Here in the Northeastern Woodlands, the traditional Wabanaki people adhere to a set of core cultural values that are contained in our sacred way of life, what we call *skejinawe bamousawakon.* These foundational teachings provide us with a solid framework upon which we can build our lives. Central among these teachings is an understanding of the deep interrelatedness of the sacred and the secular. Our traditional societies are rooted in one inseparable reality that acknowledges the inviolability of all aspects of creation. There is no separation between ceremony and our daily walk in the world. Everything is interrelated and recognized for its sacred place within the web of life. We acknowledge that the great pull of the universe is a desire to live in harmony with the Creator, which is expressed most effectively in our own lives by living harmoniously with the rest of creation.

We also recognize that time as it has been described is an illusion.

For us, time does not exist as separate epochs unfolding in linear fashion, but as one movement unfolding in all directions simultaneously. We realize that we cannot separate ourselves from those who have come before us or those who will follow, because we all exist together in this one moment. The harm experienced by our ancestors is felt in our bodies today, and the harm we create today will be experienced by our future generations tomorrow. We are all inextricably linked through these shared experiences that cross time. Our sense of responsibility for the coming generations is reflected in this awareness. It is this way of life that has allowed us to exist in a balanced relationship with our local ecosystems for more than ten thousand years. And it is this sacred way of life that can bring humanity back into alignment with a future that not only ensures human survival on Earth but also nurtures a mended relationship between human beings and all other life. The first step on this path is for us to shift the central point of our awareness.

The anthropocentric beliefs and philosophies that have ruled mainstream ideologies for generations are incapable of accommodating the holistic view needed to escape our current predicament. In my book *Sacred Instructions: Indigenous Wisdom for Living Spirit-Based Change*, I write:

> The entire span of human life exists within each one of us, going all the way back to the hands of the Creator. In our bodies, we carry the blood of our ancestors and the seeds of the future generations. We are the living conduit to all life.

This does not mean only human life but all life that has ever existed upon Mother Earth. All our creation stories teach us that we are born out of the same foundational elements that make up all life in the known universe. In one story, we enter this world when Kluskap shoots an arrow into the ash tree and creates an opening for us to emerge. The ash tree offers humankind the energy of life from within its roots to provide us passage into this world. In another story, the first human is formed from the soil of Mother Earth. In that story the eyes are the first part of the body to be created. Once the eyes are created, the first human remains in the soil for an entire cycle of life watching how the rest of creation moves, before their arms and legs are given to them.

In the story of Sky Woman, the mother of all life looks down and sees that the Earth is made entirely of water, with no land for her children to live upon. So she descends from the stars to form a landmass on the back of a turtle, creating what is now known as Turtle Island. In the story, Sky Woman enlists the help of the animals to gather soil from the bottom of the great water so that she may build the land that will sustain her children once they are born. One animal after another volunteers to dive into the water and retrieve the soil. And one by one they all return to the surface with the same story—the waters are too deep. They could not reach the bottom. Eventually, the humble *kiwhos* (muskrat) comes forward and offers to try to gather the soil. He enters the water and is gone for a very long time, and the other animals become greatly concerned for his safety. Eventually, his tiny, lifeless body floats back to the surface and the other animals bring him onto the turtle's back. As the animals are grieving the loss of their friend, one of them notices that his fist is clenched around something. When Sky Woman opens his hand, she finds the soil that she needs to form land. She is so moved by his sacrifice that she breathes life back into him so that he can live once again. From that point forward, *kiwhos* has made his home in the place where the land and water meet. Sky Woman created Turtle Island and gave birth to the Indigenous peoples of North America.

In each of these stories, the elements of the natural world play a central role in the creation of human life. These stories teach us that our life here upon Mother Earth has been supported by the beings of the natural world. They also remind us that these beings preceded us into this world, making them our elders. Through these stories, these beings became our grandmothers, grandfathers, aunts, and uncles, *N'dilnabamuk*. They became our relatives.

Science has finally confirmed that human beings share genes with all living organisms. Long before this truth was discovered by modern science, we had these stories and a sense of kinship and responsibility toward our relatives in the natural world. This one teaching has been central to the sustainable way of life that has kept Indigenous peoples in balanced harmony with our local ecosystems for tens of thousands of years. Kinship defines how we relate to one another. It determines whom and what we include in the structuring of our societies and whom and what we feel a sense of responsibility toward. If we look at our societies today, it is quite clear who and what has been excluded, and the incredible cost that this exclusion has had on our societies and our world.

One of the more damaging effects of colonization and forced assimilation has been the homogenization of our societies. Under colonial regimes, the voices of those on the fringes of society or living completely outside of the colonial system have been silenced or ignored. Within these regimes, there has also been a centralization of power and authority within a very limited demographic of society (wealthy White men), which has further limited the universe of knowledge and access to new ideas and perspectives. The stifling of creative thinking and suppression of critical analysis that results from such narrow representation erodes the overall intelligence of the entire society. These restrictive systems provide no opportunity for alternative ways of being to be presented. And since cultural ways of being are passed on from one generation to the next, there has been very little variation in the destructive patterns that brought us to this place of collective crises. The overall lack of diversity within the patriarchal colonial paradigm has had a suffocating impact on creative intelligence and a divisive impact on society.

Diversity fosters social coherence, creating more stable and harmonious relational networks, which in turn lead to more stable and harmonious societies. Additionally, the more diverse a group or community, the greater the perspectives and innovations that arise and the greater the success rate for all. Human diversity is just as critical to society as biodiversity is to an ecosystem; without it there can be no healthy functioning. The loss of diversity within mainstream systems and structures has left a fracture in our societies that must now be healed, through the purposeful and systematic inclusion of diverse voices, including the voices of the natural world, within the social dialogue.

"Every [natural] soundscape . . . generates its own unique signature, one that contains incredible amounts of information," explains Bernie Krause, and when we listen closely, it "gives us incredibly valuable tools by which to evaluate the health of a habitat across the entire spectrum of life." The voices of the natural world can inform us when an ecosystem has been unsettled by human activity. This is something that Indigenous peoples have always known, because we have been listening to the voices of the natural world in our territories for thousands of years. When we recognize the personhood of the beings in the natural world, we start to recognize that they have something valuable to teach us. The rocks, the eldest among us, carry four billion years of stories. This is why we refer to them as grandfathers: because they carry an enormous amount of diverse information and perspectives. Every plant, tree, and animal carries its own unique wisdom and can teach us how to live harmoniously with one another and in relationship with Mother Earth. When we extend our view of kinship beyond our anthropocentric view, a whole new world of knowledge becomes available to us.

Another benefit of kincentric awareness is that it informs us of our responsibilities and obligations. When we have a bond of kinship with another, it impacts the way that we care for them. In healthy systems, we treat our kin with a greater degree of care, often expressing more gentleness and protectiveness toward them. It is the same when we love. When we love, we treat the beloved with reverence and respect. Indigenous peoples living in accordance with these beliefs have lived in loving relationship with the beings in the natural world for millennia.

Indigenous kinship systems provide models of reciprocal care. We care for Mother Earth and Mother Earth cares for us.

We have words in our language that help to remind us of the balance that this relationship requires. One of those words is *mamabezu;* it means "he or she has enough." It is an acknowledgment that an individual has what they need to live their life with a sense of safety and dignity. Another word, *alabezu*, means "everyone has enough." The "everyone" envisioned in that phrase includes all the beings in the natural world. When we are contemplating the value of "enough," we recognize that *mamabezu* must always be weighed against *alabezu* to ensure that there is a balance to life.

In order to survive, we must all come to realize that we do not exist solely for the benefit and development of our individual lives as human beings. Rather, our role as human beings is to evolve into a state of interbeing with the rest of life so that we may join the universal flow that is ever moving toward harmony and balance. This is the only way that life on Mother Earth will remain viable into the future.

Indigenous peoples from across Turtle Island have several prophecies that correspond to the times we are living in. These prophecies tell of the dangers inherent in abusing Mother Earth. Rather than seeing these prophecies as portents of a determined future, we see them as sacred instructions that edify our walk upon Mother Earth. They provide echoes of our ancestors' love by offering us the guidance needed to overcome the challenges of this time and survive into the future. It is these ancient prophecies that have prepared us for this time and inform us of the dire consequences of ignoring the truths unfolding around us. It is because of this ancient knowledge that Indigenous peoples have literally been laying down their lives to stop the flow of harm that is being caused by the industrial destruction of Mother Earth, because we know that this path will inevitably alter our way of life. And, if left unchecked, it will eliminate the ability of life to exist on Mother Earth.

Today prophecies are unfolding all around us. We are all observers and participants. Our ancestors spoke of a time seven generations after first contact was made with the White man, when Mother Earth would become sick as a result of human activity. They told us that the trees would begin dying from the top down, the waters would run black with pollution, and the soil would stop providing us with food as the insects began disappearing from the Earth. In 1970 a Hopi elder from the village of Hotevilla came forward and shared these words about the future we were facing:

> Nature will speak to us with its mighty breath of wind. There will be earthquakes and floods causing great disasters, changes in the seasons and in the weather, disappearance of wildlife, and famine in different forms. There will be gradual corruption and confusion among the leaders and the people all over the world, and wars will come about like powerful winds.

All these things have been foretold, and they have all come to pass. Fortunately, the stories told about this time have not all been so dire.

There are also many that tell of a time of unification and healing. In 1877 Crazy Horse delivered a powerful prophecy following a ceremony with Sitting Bull. This ceremony took place a stone's throw from where the water protectors gathered at Standing Rock in 2016 to stop the Dakota Access Pipeline, when the Lakota people took a vital stand for the preservation of life by protecting our sacred waters. During that 2016 event, people from all corners of the Earth came to stand with the Lakota to seek their "knowledge and understanding of unity among all living things." As a result of that stand, young people around the world now carry the seed of understanding that was given to them by the Lakota people. This is something that Crazy Horse spoke of nearly 140 years earlier. Here are his words:

> Upon suffering beyond suffering, the Red Nation shall rise again and it shall be a blessing for a sick world. A world filled with broken promises, selfishness, and separations, a world longing for light again. I see a time of seven generations when all the colors of mankind will gather under the sacred Tree of Life and the whole Earth will become one circle again. In that day, there will be those among the Lakota who will carry knowledge and understanding of unity among all living things, and the young white ones will come to those of my people and ask for this wisdom. I salute the light within your eyes where the whole universe dwells. For when you are at that center within you and I am at that place within me, we shall be as one.

Today people from all corners of the world are beginning to come together to address the sickness that has overtaken the world, and they are looking to Indigenous peoples to guide them in this process.

The Anishinaabe prophecy of the seven fires provides clear guidance for non-Indigenous peoples during this crucial time. The prophecy highlights seven periods of time linked to Indigenous peoples' contact with the light-skinned people from across the water. It offers advice and warnings for each of these time periods. Ojibway-Anishinaabe elder and wisdom keeper Jim Dumont tells us that we are in the time of the seventh fire. This is what the prophecy has to say about this time:

> If the new people learned to trust the ways of the circle and trained themselves to hear their inner voice, wisdom would return to them in waking and sleeping dreams, and the sacred fire would be lit once again. Then the light-skinned people would be given a choice of two paths. If they chose the correct path, then the Seventh Fire would be used to light an Eighth Fire, which would be a lasting fire of unity and peace. If they choose the wrong path, and stay locked into their old mind-set, the destruction they wrought will come back and destroy them, and all the people of the Earth will experience great suffering and death.

We have reached the point of choice, where the light-skinned people must decide which path they will choose: take the path of unity and peace, or stay on the current path and destroy themselves and countless others with them. Whenever I hear this portion of the seven fires prophecy, it reminds me of a quote by Felix Cohen, which also seems a bit prophetic:

> Like the miner's canary, the Indian marks the shift from fresh air to poison gas in our political atmosphere; and our treatment of Indians, even more than our treatment of other minorities, reflects the rise and fall in our democratic faith.

Cohen's warning to his contemporaries that Indigenous peoples were the miner's canary within the political atmosphere can easily be extended to our current environmental catastrophes. Indigenous peoples have been sounding the alarm for centuries. In the early twentieth century, Oglala Lakota chief John Hollow Horn told the U.S. government, "Some day the Earth will weep, she will beg for her life, she will cry with tears of blood. You will make a choice, if you will help her or let her die, and when she dies, you too will die." One hundred years earlier, my Wabanaki ancestors, Passamaquoddy Nation delegates Deacon Sockabasin and Joseph Stanislaus, stood before the Maine Legislature asking that the state stop the destruction of the rivers and forests. Today Indigenous peoples across the Americas are literally placing their bodies in harm's way to stop the destruction of Mother Earth. I often wonder when the rest of the world will finally hear us.

When will the mainstream population finally realize that the annihilation of Indigenous peoples is also the annihilation of humankind on Mother Earth? For how can humankind continue to live when the keepers of the umbilical connection to Mother Earth have been destroyed? Who will nurture that connection when we are gone?

The Indigenous way of life is a pathway that can lead humankind back toward life. It provides a way of being that is in harmony and balance with the rest of creation. It reconnects humankind to the sources of its survival and the heart of its humanity. This way of life is a gift that was given to us by our ancestors. And it can be a gift that we collectively offer to our future if we have the courage to walk it together. If we listen closely, we can hear the voices of our ancestors encouraging us to follow this path. As writer Linda Hogan imagines:

> Be still, they say. Watch and listen. You are the result of the love of thousands.

Among those thousands of ancestors are the ash tree, who generously offered her life force to transport us into this world; the soil of Mother Earth, who lovingly held the first human being as they were being formed; and the humble muskrat, who selflessly sacrificed his life so that we would have the privilege of living upon this land. Once we restore these relatives to their rightful place within our kinship networks, the entire lineage of life will rise up to meet us and we will begin moving back into balance with the flow of the universe and solidly in the direction of renewing Mother Earth. *Psilde N'dilnabamuk*—I offer this for all my relations.

Soul and soil are not separate. Neither are wind and spirit, nor water and tears. We are eroding and evolving, at once, like the red rock landscape before me. Our grief is our love. Our love will be our undoing as we quietly disengage from the collective madness of the patriarchal mind that says aggression is the way forward.

—Terry Tempest Williams

A Handful of Dust

Kate Marvel

It rains in the Amazon because the trees want it to. There is plenty of moisture in the oceans that surround the continent, but there is also a hidden reservoir on the land, feeding an invisible river that flows upward to the sky. The water held in the soil is lifted up by the bodies of the trees and lost to the atmosphere through the surfaces of their leaves. The local sky plumps with moisture, primed for the arrival of the seasonal rains driven by the annual back-and-forth march of the sun's rays. As climate scientist Alex Hall puts it, the trees are co-conspiring with the sky to attract an earlier monsoon.

One night in the future, a jet takes off from an obscure equatorial airstrip deep in the Amazon. It is an ordinary Gulfstream, an unassuming workhorse for the world's wealthy. Today, though, there are no passengers aboard; it carries only secrets. From the windows of the jet, the pilot cannot see the forests below, only a blue sky interrupted by a white patchwork, alternately thick and wispy, that hides every hint of green and brown. But it is those forests, and the people who live in them, and the people far afield who may breathe in oxygen they breathe out, that the pilot is trying to save.

The forests are dying, attacked on all sides by relentless demand for fossil fuels, beef, money. Humans start fires; as the temperature warms, every season becomes fire season. As the atmosphere is larded with carbon dioxide, the plants of the Amazon do not need to open the pores on their leaves quite so much to take in the gases they need. These shrunken pores expel less water into the atmosphere. The trees are losing their ability to summon the monsoon, slowly becoming decoupled from the surrounding air, a forest dissolving into dust.

Miles above the ground, the jet jettisons its cargo—mineral sunscreen to be injected into the swirling air currents. The tiny particles will be smeared on the stratosphere, a protective aerosol shield that will block a little sunlight that would otherwise warm the planet. It is a Hail Mary, a desperate attempt to cool the planet by blocking the sun. They call it geoengineering—*geo*, the Greek for "earth," *engineer* from the Middle English for "contrive, deceive, put to torture." It is not a good idea. But it has happened before.

It is still dark when the plane returns to the airstrip, but in the darkness the lights of this jet, and others returning from the same mission, shine brightly. The next morning, the sun rises in a blaze of brilliant red. It is the most beautiful sunrise anyone can remember. It lasts only a few minutes before the rains come again.

The year is 1816, and there will be no summer. In New England and Canada, brutal May frosts kill the crops. A housewife notes in her diary, simply, "Weather backward." The Indian Ocean monsoon, usually triggered by warm summer temperatures, is delayed by months. When it finally comes to the subcontinent, the torrential rains drown croplands and the people who work them, leaving pools of stagnant, filthy water. Cholera spreads as far as Moscow. Harvests fail in northern Europe, fraying societies battered by the Napoleonic Wars. Switzerland—*Switzerland*—is racked by violence as desperate, starving mobs converge on the cities.

What history most remembers from the carnage spreading across the Northern Hemisphere that year is the aftermath of a bad vacation. In a villa on Lake Geneva, a group of bored English tourists challenge one another to write ghost stories. It is cold; nearly every day of June and July has been rainy. The gloom—dulled, perhaps, by heavy opium use and sexual tension—inspires Lord Byron and his doctor and frenemy, John William Polidori, to create the outlines of the vampire tale that would become *Dracula*. But it is the novel written by the teenage girl trapped inside with these insufferable men that will become most known to posterity. The soggy, relentless weather gave birth to Mary Shelley's *Frankenstein* and his tragic, violent, misunderstood monster.

Is it any wonder that we scientists, tarred for eternity by this portrayal of hubris and vanity, are curious to understand the circumstances under which it came about?

The culprit—suppressor of summer, bringer of rain, gothic muse—is to be found on Sumbawa, one of the Lesser Sunda Islands in what was then the Dutch East Indies and is now Indonesia. On the Sanggar Peninsula, among cashew plantations and a few scattered small villages, you can see the remains of what was once a perfectly formed volcanic cone. In April of 1815, Mount Tambora erupted, blasting gas and dust high into the stratosphere and directly killing over ten thousand islanders. It was by far the most destructive explosion in human history.

Few Europeans would have known of the eruption. Information traveled slowly in those early-industrial times, and the news of an obscure volcano blowing up an out-of-the-way colony would not have been of general interest. But the violence of the event, and its tropical location, meant that its ejecta were very effectively injected into the upper atmosphere. From there, stratospheric winds took over, dispersing the fog of volcanic gas and dust to far-off skies.

Frankenstein was a scientist. The monster was nameless. It never existed, except as a metaphor for curiosity turned to hubris and then tragedy. The future, with its sunscreen-smearing jets and its desperate attempts to cool the planet, is presumably imaginary—for now.

But today, from the safety of our climate-controlled offices, scientists can set off multiple Tamboras. A few lines of code, and the volcano is doubled in size or moved halfway around the world. We can make it explode millions of years ago or in the next decade. We've built toy planets—climate models that live on supercomputers—and we can manipulate them like malevolent gods. We can run experiments on the entire Earth and repeat them over and over until we've learned something. By setting off volcanoes in these models, we now know the conditions under which they can cool the planet, and what happens after they do.

And we've learned another thing: Tambora was not unique. Plot the average temperature of the globe, and the graph's rising line will be punctuated by the occasional sharp downward spike. Krakatoa in 1883, then the twentieth-century eruptions of Santa María, Agung, El Chichón, and finally Pinatubo in 1991: five explosions since Tambora powerful enough to spray gas and dust far up into the stratosphere. The sun-blocking effects were powerful—Pinatubo cooled the planet by an entire degree—but short-lived. Eventually the atmosphere cleansed itself, and the protective shield of tiny particles disintegrated and fell to earth.

Imagine, though, if volcanoes exploded like clockwork, one every few years, spewing a regular injection of gas and dust high into the stratosphere. Imagine controlled explosions that require no mountain, no cinder cone, no flow of lava onto farmlands below. Imagine, in short, if *we* were the volcano.

Courage is the resolve to do well without the assurance of a happy ending. But even in desperate times, there is a line between brave and foolish. We have to do *something*. It does not follow that we should simply settle for *anything*.

Everything is connected, and everything is complicated. We do not live on a planet governed by a single variable. The crops that feed us, the reservoirs that slake our thirst, and the rivers that transport our goods depend on the amount and timing of rainfall. The speed and direction of prevailing winds dictate everything from the position of our wind turbines to the routes taken by transatlantic jets. The planet is connected to itself in strange and surprising ways, and we still have only a limited understanding of these intricate links.

The climate, we are told, is always changing. We know this. Volcanic eruptions cancel summers and freeze early-season fruit on the vine. The sun's output varies slightly. Wobbles in the Earth's orbit precipitate ice ages and deglaciations. The changes that have occurred before—Tambora, Pinatubo, the glaciers that covered much of Europe and North America twenty thousand years ago—show the fragility of the planet and its capacity for change. These changes are different from the changes we are now effecting, but they tell us things, and we live with their consequences even now.

There were once forests in the Sahara—if not quite the Amazon, still lush and tropical, clustered around the largest freshwater lake on the planet at the time. The lake is mostly gone now, vanished in the span of a few hundred years. In its place there is nothing but dust. It is dry because the planet wobbled slightly in its orbit, weakening the monsoon rains in the west of Africa. The plants sucked the moisture from the soil; it was not replaced. They died, and no more moisture entered the atmosphere: a vicious cycle of dying and drying that led to the depopulated desert we know today. The climate changed; it was not the fault of humans. But the existence of past climate change does not mean we are not responsible for it this time. There have always been gentle and natural deaths. This does not make murder impossible.

There is no death without life. There are no deserts without the tropics. The winds over the Sahara come from the east—dense, sinking, forced sideways as the Earth rotates away underneath the air. Saharan dust is carried across the Atlantic, enlarging the beaches of the Caribbean, scattering low-angle sunlight into brilliant purple-orange sunsets, and landing gently on the forests of the Amazon. The air over the Sahara has itself arrived from the tropics, rising and shedding its moisture on a journey toward the poles. When it can go no farther, it cools and sinks, pressing its dry weight down onto the parched land below.

The desert gives back to the tropics. The Amazonian rainforest is so lush that it cannot fertilize itself. Nutrients are seized by the greedy vegetation, locked up in the bodies of plants before they can leach into the soil. But the forest *is* fertilized, given life by the dead lake in the Sahara. There is phosphorus in the old lake bed, swept up and across the Atlantic by the prevailing winds. The Sahara's dried-up lake retains a chemical memory of what it used to be. And it is this that helps to create the Amazon.

The living Earth is a sum of delicate balances, the culmination of a more-than-four-billion-year history of improbable coincidences and opportunistic alliances. It is this system that humans have perturbed by changing the chemistry of the atmosphere with our excesses of emissions. And it is this patchwork of miracles that the jets, laden with stratospheric sunscreen, are purportedly trying to preserve. Whether you call it geoengineering, solar-radiation management, or artificial volcanism, they are countering unthinking, carbon-propelled climate change with deliberate, managed climate change and naively hoping for a cancellation.

It is easy to imagine but difficult to quantify the possible horrors triggered by particles in the stratosphere. The monsoon comes late to the Indian subcontinent, or not at all. In one poor country the rivers run dry; in another they burst their banks. The jet stream becomes erratic and unpredictable. The weather in England turns Siberian. Canada melts.

All the caveats, all the hesitation we scientists invoke when projecting specifics of future climate—we do so because even with all we know, much remains beyond our predictive capacities. How could the solution be *more* man-made climate change, not less? And that is geoengineering's premise: yet more monsters of our own making. Dominoes will fall in a rapidly accelerating chain, but how significant and where they might be is a mystery. It would have been safer never to have touched the first one.

Because: The universe tends to disorder. Things fall apart. Perfume spritzed into the air, cigarette smoke in a bar, ricin released on a crowded subway: The minute the button is pushed, something irreversible is set in motion. The consequences of tiny, random acts echo throughout the world. Jets can disperse sunblock throughout the stratosphere, but no fleet is large or sophisticated enough to recapture every particle once they escape. There is no fallback mode, nothing to do but wait until the air cleanses itself. If managing the sun is a terrible mistake, if that act leads to starvation or war or collapse, it is an irreversible one.

We don't know everything. But we don't know nothing. Even amid the uncertainty, we can pick out clear threads. We are inevitably sending our children to live on an unfamiliar planet. But if the future is to be unmitigated greenhouse gases countered by sun-blocking particles, if we mask the hangover but continue the bender, we will have to be resigned to losing even more of the world we know. Imagine: The oceans will still choke on carbon dioxide, turning inexorably to acid. Corals and sea creatures will be lost. Less sunlight will be available for plants to grow. Sulfur dioxide in the stratosphere will rip a new hole in the ozone layer, reversing its slow healing process. Without this protective layer, UV rays will reach our skins, giving more of us cancer. Spewed dust and particles and gases will parry and scatter the fading light of dusk. All of this we know for certain.

We'll have spectacular sunsets. But during the day, the light will seem weak and white and hazy. Because this is the trade-off: If we do this, the sky will never be as blue again.

November

Lynna Odel

If I can't save us

then let me feel you
happy and safe
under my chin.

If this will drown
or burn

then let us drink starlight
nap under trees
sing on beaches—

the morning rush to sit indoors is for
what, again?

If we are dying

then let me rip open
and bleed Love,
spill it, spend it
see how much
there is

the reward for misers is
what, again?

If this life is ending

then let me begin
a new one

What Is Emergent Strategy?

ADRIENNE MAREE BROWN

"Emergence is the way complex systems and patterns arise out of a multiplicity of relatively simple interactions"—these words from Nick Obolensky are the clearest articulation of emergence that I have come across. In the framework of emergence, the whole is a mirror of the parts. Existence is fractal—the health of the cell is the health of the species and the planet.

There are examples of emergence everywhere.

Birds don't make a plan to migrate, raising resources to fund their way, packing for scarce times, mapping out their pit stops. They feel a call in their bodies that they must go, and they follow it, responding to each other, each bringing their adaptations.

There is an art to flocking: staying separate enough not to crowd each other, aligned enough to maintain a shared direction, and cohesive enough to always move toward each other. (Responding to destiny together.) Destiny is a calling that creates a beautiful journey.

Emergence is beyond what the sum of its parts could even imagine.

A group of caterpillars or nymphs might not see flight in their future, but it's inevitable.

It's destiny.

Oak trees don't set an intention to listen to each other better or agree to hold tight to each other when the next storm comes. Under the earth, always, they reach for each other, they grow such that their roots are intertwined and create a system of strength that is as resilient on a sunny day as it is in a hurricane.

Dandelions don't know whether they are a weed or a brilliance. But each seed can create a field of dandelions. We are invited to be that prolific. And to return fertility to the soil around us.

Cells may not know civilization is possible. They don't amass as many units as they can sign up to be the same. No—they grow until they split, complexify. Then they interact and intersect and discover their purpose—*I am a lung cell! I am a tongue cell!*—and they serve it.

And they die. And what emerges from these cycles are complex organisms, systems, movements, societies.

Nothing is wasted, or a failure. Emergence is a system that makes use of everything in the interactive process. It's all data.

Octavia Butler said, "Civilization is to groups what intelligence is to individuals. It is a means of combining the intelligence of many to achieve ongoing group adaptation."

She also said, "All that you touch you change / all that you change, changes you." We are constantly impacting and changing our civilization—each other, ourselves, intimates, strangers. And we are working to transform a world that is, by its very nature, in a constant state of change.

Janine Benyus, a student of biomimicry, says: Nature/Life would always "create conditions conducive to life." She tells of a radical fringe of scientists who are realizing that natural selection isn't individual but mutual—that species survive only if they learn to be in community.

How can we, future ancestors, align ourselves with the most resilient practices of emergence as a species?

Many of us have been socialized to understand that constant growth, violent competition, and critical mass are the ways to create change. But emergence shows us that adaptation and evolution depend more upon critical, deep, and authentic connections, a thread that can be tugged for support and resilience. The quality of connection between the nodes in the patterns.

Dare I say love.

And we know how to connect—we long for it.

On Fire

Naomi Klein

It has been over three decades since governments and scientists started officially meeting to discuss the need to lower greenhouse gas emissions to avoid the dangers of climate breakdown. In the intervening years, we have heard countless appeals for action that involve "the children," "the grandchildren," and "generations to come." We were told that we owed it to them to move swiftly and embrace change. We were warned that we were failing in our most sacred duty to protect them. It was predicted that they would judge us harshly if we failed to act on their behalf.

Well, none of those emotional pleas proved at all persuasive, at least not to the politicians and their corporate underwriters who could have taken bold action to stop the climate disruption we are all living through today. Instead, since those government meetings began in 1988, global carbon dioxide emissions have risen by well over 40 percent, and they continue to rise. The planet has warmed by about 1 degree Celsius since we began burning coal on an industrial scale, and average temperatures are on track to rise by as much as four times that amount before the century is up; the last time there was this much carbon dioxide in the atmosphere, humans didn't exist.

As for those children and grandchildren and generations to come who were invoked so promiscuously? They are no longer mere rhetorical devices. They are now speaking (and screaming, and striking) for themselves. And they are speaking up for one another as part of an emerging international movement of children and a global web of creation that includes all those amazing animals and natural wonders that they fell in love with so effortlessly, only to discover that it was all slipping away.

And yes, as foretold, these children are ready to deliver their moral verdict on the people and institutions who knew all about the dangerous, depleted world they would inherit and yet chose not to act.

They know what they think of Donald Trump in the United States and Jair Bolsonaro in Brazil and Scott Morrison in Australia and all the other leaders who torch the planet with defiant glee while denying science so basic that these kids could grasp it easily at age eight. Their verdict is just as damning, if not more so, for the leaders who deliver passionate and moving speeches about the imperative to respect the Paris Climate Agreement and "make the planet great again" (France's Emmanuel Macron, Canada's Justin Trudeau, and so many others) but who then shower subsidies, handouts, and licenses on the fossil fuel and agribusiness giants driving ecological breakdown.

Young people around the world are cracking open the heart of the climate crisis, speaking of a deep longing for a future they thought they had but that is disappearing with each day that adults fail to act on the reality that we are in an emergency.

This is the power of the youth climate movement. Unlike so many adults in positions of authority, they have not yet been trained to mask the unfathomable stakes of our moment in the language of bureaucracy and overcomplexity. They understand that they are fighting for the fundamental right to live full lives—lives in which they are not, as fourteen-year-old climate striker Alexandria Villaseñor puts it, "running from disasters."

For the first global youth strike, on March 15, 2019, organizers estimate there were nearly 2,100 strikes in 125 countries, with 1.6 million young people participating. That's quite an achievement for a movement that began just eight months earlier with a single fifteen-year-old girl in Stockholm, Sweden, Greta Thunberg.

During normal, nonemergency times, the capacity of the human mind to rationalize, to compartmentalize, and to be distracted easily is an important coping mechanism. All three of these mental tricks help us get through the day. It's also extremely helpful to look unconsciously to our peers and role models to figure out how to feel and act—those social cues are how we form friendships and build cohesive communities.

When it comes to rising to the reality of climate breakdown, however, these traits are proving to be our collective undoing. They are reassuring us when we should not be reassured. They are distracting us when we should not be distracted. And they are easing our consciences when our consciences should not be eased.

In part this is because if we were to decide to take climate disruption seriously, pretty much every aspect of our economy would have to change, and there are many powerful interests that like things as they are. Not least the fossil fuel corporations, which have funded a decades-long campaign of disinformation, obfuscation, and straight-up lies about the reality of global warming.

As a result, when most of us look around for social confirmation of what our hearts and heads are telling us about climate disruption, we are confronted with all kinds of contradictory signals, telling us instead not to worry, that it's an exaggeration, that there are countless more important problems, countless shinier objects to focus on, that we'll never make a difference anyway, and so on. And it most certainly doesn't help that we are trying to navigate this civilizational crisis at a moment when some of the most brilliant minds of our time are devoting vast energies to figuring out ever-more-ingenious tools to keep us running around in digital circles in search of the next dopamine hit.

This may explain the odd space that the climate crisis occupies in the public imagination, even among those of us who are actively terrified of climate collapse. One minute we're sharing articles about the insect apocalypse and viral videos of walruses falling off cliffs because sea-ice loss has destroyed their habitat, and the next we're online shopping and willfully turning our minds into Swiss cheese by scrolling through Twitter or Instagram. Or else we're binge-watching Netflix shows about the zombie apocalypse that turn our terrors into entertainment, while tacitly confirming that the future ends in collapse anyway, so why bother trying to stop the inevitable? It also might explain the way serious people can simultaneously grasp how close we are to an irreversible tipping point and still regard the only people who are calling for this to be treated as an emergency as unserious and unrealistic.

"I think in many ways that we autistic are the normal ones, and the rest of the people are pretty strange," Thunberg has said, adding that it helps not to be easily distracted or reassured by rationalizations. "Because if the emissions have to stop, then we must stop the emissions. To me that is black or white. There are no gray areas when it comes to survival. Either we go on as a civilization or we don't. We have to change." Living with autism is anything but easy—for most people, it "is an endless fight against schools, workplaces and bullies. But under the right circumstances, given the right adjustments it can be a superpower."

The wave of youth mobilization that burst onto the scene in March 2019 is not the result of one girl and her unique way of seeing the world. Greta is quick to note that she was inspired by another group of teenagers who rose up against a different kind of failure to protect their futures: the students in Parkland, Florida, who led a national wave of class walkouts demanding tough controls on gun ownership after seventeen people were murdered at their school in February 2018.

Nor is Thunberg the first person with tremendous moral clarity to yell "Fire!" in the face of the climate crisis. It has happened multiple times over the past several decades; indeed, it is something of a ritual at the annual UN summits on climate change. But perhaps because these earlier voices belonged to Brown and Black people from the Philippines, the Marshall Islands, and South Sudan, those clarion calls were one-day stories, if that. Thunberg is also quick to point out that the climate strikes themselves were the work of thousands of diverse student leaders, their teachers, and supporting organizations, many of whom had been raising the climate alarm for years.

As a manifesto put out by British climate strikers put it, "Greta Thunberg may have been the spark, but we're the wildfire."

As deep as our crisis runs, something equally deep is also shifting, and with a speed that startles me. As I write these words, it is not only our planet that is on fire. So are social movements rising up to declare, from below, a people's emergency. In addition to the wildfire of student strikes, we have seen the rise of Extinction Rebellion, which exploded onto the scene and kicked off a wave of nonviolent direct action and civil disobedience, including a mass shutdown of large parts of central London. Extinction Rebellion is calling on governments to treat climate change as an emergency, to rapidly transition to 100 percent renewable energy in line with climate science, and to democratically develop a plan for how to implement that transition through citizens' assemblies. Within days of its most dramatic actions in April 2019, Wales and Scotland both declared a state of climate emergency, and the British parliament, under pressure from opposition parties, quickly followed suit.

In this same period in the United States, we have seen the meteoric rise of the Sunrise Movement, which burst onto the political stage when it occupied the office of Nancy Pelosi, the most powerful Democrat in Washington, DC, one week after her party had won back the House of Representatives in the 2018 midterm elections. Wasting no time on congratulations, the Sunrisers accused the party of having no plan to respond to the climate emergency. They called on Congress to immediately adopt a rapid decarbonization framework, one as ambitious in speed and scope as Franklin D. Roosevelt's New Deal, the sweeping package of policies designed to battle the poverty of the Great Depression and the ecological collapse of the Dust Bowl.

The activism we are seeing today builds on this history and also changes the equation completely. Though many of the efforts just described were large, they still engaged primarily with self-identified environmentalists and climate activists. If they did reach beyond those circles, the engagement was rarely sustained for more than a single march or pipeline fight. Outside the climate movement, there was still a way that the planetary crisis could be forgotten for months on end or go barely mentioned during pivotal election campaigns.

Our current moment is markedly different, and the reason for that is twofold: one part having to do with a mounting sense of peril, the other with a new and unfamiliar sense of promise.

One month before the Sunrisers occupied the office of soon-to-be House Speaker Nancy Pelosi, the UN Intergovernmental Panel on Climate Change (IPCC) published a report that had a greater impact than any publication in the thirty-one-year history of the Nobel Peace Prize–winning organization.

The report examined the implications of keeping the increase in planetary warming below 1.5 degrees Celsius (2.7 degrees Fahrenheit). Given the worsening disasters we are already seeing with warming of about 1 degree Celsius, it found that keeping temperatures below the 1.5-degree threshold is humanity's best chance of avoiding truly catastrophic unraveling.

But doing that would be extremely difficult. According to the

World Meteorological Organization, we are on a path to warming the world by 3 to 5 degrees Celsius by the end of the century. Turning our economic ship around in time to keep the warming below 1.5 degrees Celsius would require, the IPCC authors found, cutting global emissions approximately in half by 2030 and getting to net-zero carbon dioxide emissions by 2050. Not just in one country but in every major economy. And because carbon dioxide in the atmosphere has already dramatically surpassed safe levels, it would also require drawing a great deal of that down, whether through unproven and expensive carbon capture technologies or the old-fashioned ways: restoring forests and other ecosystems, and farming in ways that regenerate soil.

Pulling off this high-speed pollution phaseout, the report establishes, is not possible with singular technocratic approaches like carbon taxes, though those tools must play a part. Rather, it requires deliberately and immediately changing how our societies produce energy, how we grow our food, how we move ourselves around, and how our buildings are constructed. What is needed, the report's summary states in its first sentence, is "rapid, far-reaching and unprecedented changes in all aspects of society."

This was not the first terrifying climate report by any means, nor the first unequivocal call from respected scientists for radical emissions reduction. My bookshelves are crowded with these findings. But like Greta Thunberg's speeches, the starkness of the IPCC's call for root-and-branch societal change, and the shortness of the time line it laid out for pulling it off, focused the public mind like nothing before.

It was against this backdrop that 2019's cascade of large and militant climate mobilizations unfolded. Again and again at the strikes and protests, we heard the words "We have only twelve years." Thanks to the IPCC's unequivocal clarity, as well as direct and repeated experience with unprecedented weather, our conceptualization of this crisis is shifting. Many more people are beginning to grasp that the fight is not for some abstraction called "the Earth." We are fighting for our lives.

As powerful a motivator as the IPCC report has proven to be, perhaps an even more important factor has to do with the calls coming from many different quarters in the United States and around the world for governments to respond to the climate crisis with a sweeping Green New Deal. The idea is a simple one: In the process of transforming the infrastructure of our societies at the speed and scale that scientists have called for, humanity has a once-in-a-century chance to fix an economic model that is failing the majority of people on multiple fronts—because the factors that are destroying our planet are also destroying people's quality of life in many other ways, from wage stagnation to gaping inequalities to crumbling services to the breakdown of any semblance of social cohesion. Challenging these underlying forces is an opportunity to solve several interlocking crises at once.

The various plans that have emerged for a Green New Deal–style transformation envision a future where the difficult work of transition has been embraced, including sacrifices in profligate consumption. But in exchange, day-to-day life for working people has been improved in countless ways, with more time for leisure and art, truly accessible and affordable public transit and housing, yawning racial and gender wealth gaps closed at last, and city life that is not an unending battle against traffic, noise, and pollution.

Long before the IPCC's 1.5-degrees report, the climate movement had focused on the perilous future we faced if politicians failed to act. We popularized and shared the latest terrifying science. We said no to new oil pipelines, gas fields, and coal mines; no to universities, local governments, and unions investing endowments and pensions in the companies behind these projects; no to politicians who denied climate change and no to politicians who said all the right things but did the wrong ones. All this was critical work, and it remains so. But while we raised the alarm, only the relatively small "climate justice" wing of the movement focused its attention on the kind of economy and society we wanted instead.

That was the game changer of the Green New Deal bursting into the political debate in November 2018. Wearing shirts that read "We have a right to good jobs and a livable future," hundreds of young members of the Sunrise Movement chanted for a Green New Deal as they lined the halls of Congress shortly after the 2018 midterms. There was finally a big and bold "yes" to pair with the climate movement's many "no"s, a story of what the world could look like after we embraced deep transformation, and a plan for how to get there.

If the IPCC report was the clanging fire alarm that grabbed the attention of the world, the Green New Deal is the beginning of a fire-safety and prevention plan, not a piecemeal approach that merely trains a water gun on a blazing fire, as we have seen so many times in the past, but a comprehensive and holistic plan to actually put out the fire. Especially if the idea spreads around the world—which is already beginning to happen.

Those of us who advocate for this kind of transformative platform are sometimes accused of using the climate crisis to advance a socialist or anticapitalist agenda that predates our focus on the climate crisis. My response is a simple one. For my entire adult life, I have been involved in movements confronting the myriad ways that our current economic systems grind up people's lives and landscapes in the ruthless pursuit of profit. My first book, *No Logo*, published in 2000, documented the human and ecological costs of corporate globalization, from the sweatshops of Indonesia to the oil fields of the Niger Delta. I have seen teenage girls treated like machines to make our machines and seen mountains and forests turned to trash heaps to get at the oil, coal, and metals beneath.

The painful, even lethal impacts of these practices were impossible to deny; it was simply argued that they were the necessary costs of a system that was creating so much wealth that the benefits would eventually trickle down to improve the lives of nearly everyone on the planet. What has happened instead is that the indifference to life that was expressed in the exploitation of individual workers on factory floors and in the decimation of individual mountains and rivers has trickled up to swallow our entire planet, turning fertile lands into salt flats, beautiful islands into rubble, and draining once vibrant reefs of their life and color.

I freely admit that I do not see the climate crisis as separable from the more localized market-generated crises that I have documented over the years; what is different is the scale and scope of the tragedy, with humanity's one and only home now hanging in the balance. I have always had a tremendous sense of urgency about the need to shift to a dramatically more humane economic model. But there is a different quality to that urgency now because it just so happens that we are all alive at the last possible moment when changing course can mean saving lives on a truly unimaginable scale.

None of this means that every climate policy must dismantle capitalism or else it should be dismissed (as some critics have absurdly claimed)—we need every action possible to bring down emissions, and we need them now. But it does mean, as the IPCC has so forcefully confirmed, that we will not get the job done unless we are willing to embrace systemic economic and social change.

That does not mean we simply need a New Deal painted green or a Marshall Plan with solar panels. We need changes of a different quality and character. We need wind and solar power that is distributed and, where possible, community owned, rather than the New Deal's highly centralized, monopolistic river-damming hydro and fossil fuel power. We need beautifully designed, racially integrated, zero-carbon urban housing, built with democratic input from communities of color—rather than the sprawling White suburbs and racially segregated urban housing projects of the postwar period. We need to devolve power and resources to Indigenous communities, smallholder farmers, ranchers, and sustainable-fishing folk so they can lead a process of planting billions of trees, rehabilitating wetlands, and renewing soil—rather than handing over control of conservation to the military and federal agencies, as was overwhelmingly the case with the New Deal's Civilian Conservation Corps.

And even as we insist on naming an emergency as an emergency, we need to constantly guard against this state of emergency becoming a state of exception, in which powerful interests exploit public fear and panic to roll back hard-won rights and steamroll profitable false solutions.

The message coming from the school strikes is that a great many young people are ready for this kind of deep change. They know all too well that the sixth mass extinction is not the only crisis they have inherited. They are also growing up in the rubble of market euphoria, in which the dreams of endlessly rising living standards have given way to rampant austerity and economic insecurity. And techno-utopianism, which imagined a frictionless future of limitless connection and community, has morphed into addiction to the algorithms of envy, relentless corporate surveillance, and spiraling online misogyny and White supremacy.

"Once you have done your homework," Greta Thunberg says, "you realize that we need new politics. We need a new economics, where everything is based on our rapidly declining and extremely limited carbon budget. But that is not enough. We need a whole new way of thinking. . . . We must stop competing with each other. We need to start cooperating and sharing the remaining resources of this planet in a fair way."

Because our house is on fire, and this should come as no surprise. Built on false promises, discounted futures, and sacrificial people, it was rigged to blow from the start. It's too late to save all our stuff, but we can still save one another and a great many other species too. Let's put out the flames and build something different in its place. Something a little less ornate, but with room for all those who need shelter and care.

Let's forge a Global Green New Deal—for everyone this time.

2. ADVOCATE

MADELEINE JUBILEE SAITO

Litigating in a Time of Crisis

Abigail Dillen

Watching climate change progress is like living in a dream where running is required but the reflexes are gone. Even as life goes on in familiar ways, the familiar has become unreliable. This isn't the acute alarm that comes with a new fire season in my home state of California. It's an everyday sense that I will look back and long for the peace of ironing on this sunny afternoon, for reading aloud with my boy in this safe, warm bed. It's swimming out into a calm bay on a bluebird day and feeling that under the sparkling surface, the ocean may be dying.

Over the nine months that my mother suffered and ultimately died from ALS, all my times with her were marked by a shadow awareness that each moment might be the last of its kind. Often it was. More than two years after her death, I am still learning to live with her absence. As I get older, the cumulative loss of beloved people bears down on me, but I assume I am built to withstand it. Our climate emergency is different. How can we bear this anxiety and grief for the whole world? More important, how can we find our power to act and limit the damage?

If you are feeling as alive as you have ever felt, relentlessly anxious, and incapable of being at peace, I am with you.

There are various explanations for our tragic failures to act in the face of climate change—and I mean tragic in the classical Greek sense of being brought down by our own human flaws, not by fate. Even at this late stage, we could win the race against disaster. Clean energy, zero-emissions transportation and buildings, climate-friendly agriculture—it's all plausible and under way. If we make investments on the scale that is required in the 2020s, we can not only avoid the worst of climate change but also make our lives so much better. To live in the United States, the wealthiest country in the world, and watch a steady decline in equality and happiness is to know we can do better.

I hear many reasons why we are nonetheless doomed. Most commonly, I hear that climate solutions will be too expensive, and even if the United States acts, other countries won't. Putting aside the fact that clean electricity enjoys a growing cost advantage over fossil fuels, what can "too expensive" mean when we are contemplating the inundation of the world's most populous cities, dire food and water shortages, mass migrations, the sixth extinction? And in the face of existential threats, why does the narrative of U.S. leadership, flawed as it is, break down—especially among the people most prone to invoke it? Every government has the same imperative to address climate change, and recent history confirms that U.S. engagement and investment are critical to raising ambition and spurring action globally.

More compellingly, I hear that people aren't wired to stop slow-moving disasters. The scope and scale of the problem are too overwhelming for political systems that are built to produce incremental changes. Our government can't function in the public interest because of money in politics and the concentration of wealth in a very few hands. I don't dismiss any of these concerns, especially the last. But I reject the lazy fatalism. I look at the unique positioning of the United States—vast resources to address the crisis but apparently unshakable opposition to conceding its existence—and I see White privilege driving both climate denial and the political apathy and disengagement that enable it.

Unsurprisingly, support for climate action correlates with the perception of climate risk, as well as egalitarian values. White people, and especially White men, perceive the risk least. For many people who have been comfortable until now, it is seemingly impossible to accept that anything so disruptive as climate change is possible. Either it's a hoax or things will somehow continue to be fine. It's a profound kind of entitlement to believe, despite all evidence to the contrary, that we can go on living as we are.

I don't exempt myself from this critique. I came late to an interest in power and the mechanics of social change. I grew up believing in the myth of U.S. exceptionalism. I had the impression that progress was somehow self-executing, that our system inherently guaranteed the fundamentals, like human rights, clean air and water, the pursuit of happiness. I was oblivious to structural racism and the unearned structural advantages I enjoyed. I thought my job was to hone my strengths within a system that could be trusted to work for everyone. Given the opportunity to go to schools that remain bastions of privilege, I came away with a nineteenth-century skill set and a worldview steeped in the so-called Western tradition. It took me too long to reckon with the true history of our country, and too long to realize that we bend toward justice *only* as a result of tireless struggle.

That said, the struggle is welcoming to latecomers. Living in this time of crisis, for our democracy as well as the climate, is breaking my heart open and creating space for new understanding. The weight of history is on our shoulders, but this moment is alive with possibility.

As our climate deadlines loom and I feel compelled to do more, I think of my mom, with her practical handle on justice and her tireless knack for building community.

She was a trial lawyer in the days when even fewer women were. In the early days of her career, she would often appear in court in a silk dress, a signature rebellion against the expectation that she be as little like a woman as possible. She would go deep to understand why her clients had been injured, and woe to the evasive wrongdoer. She was interested in the truth, and she was restless and maddeningly single-minded until she was satisfied she knew it. She loved to go over the whole story of a case at the dinner table, underscoring facts and people that would become important later, constructing her argument with evenly laid bricks of evidence and the mortar of genuine moral outrage, for which she had a special gift.

I watched her deliver a closing argument when I was eleven or twelve. I don't remember what the case was about, although I'm sure I knew the particulars at the time. What I do remember vividly are the expressions and the postures of the jurors, the intensity of their attention, how purposeful and even eager they seemed to *go* with her. In retrospect, I think it was always vindicating to hear her speak as plainly as she did about unfairness. You had the feeling you were dealing with someone who could and would deliver justice.

I understand why courtroom dramas are evergreen favorites. Unfair as the courts and our laws can be, there is no institution in our society that is more expressly designed to privilege truth over power. My mother used the power of law to help people in ways that forever altered the course of their lives for the better. Even outside the courtroom, she was addicted to people and problem solving. It expanded the reach and impact of her deeply personal commitments—to helping other women, to children and their welfare, to education and access to the courts, to her community. I don't think she thought consciously about inclusion, but she fostered it. At her hastily planned memorial, there wasn't enough room for the thousand or more people who showed up on a Tuesday morning to pay their respects.

Somewhere behind my unconsidered decision to go to law school was my mother. But at the time, I was sure I was just having a failure of imagination. Law school did not give my life purpose, but a summer clerkship at Earthjustice did. Twenty-one years later, I look back on that summer in Montana and realize it was my first brush with real power.

As the climate emergency schools us anew, everything—our health, our wealth, our struggles for justice, the web of life on Earth—*everything* depends on what we blandly call "the environment." Relegated to the status of one "issue" among many in the United States, the environment has languished as a second-tier concern for most of my lifetime. But I was born during a phenomenal burst of consciousness and lawmaking, and I have come to cherish our bedrock environmental laws as a birthright.

By the close of the 1960s, the environment was in bad enough shape to prompt a political response. The Cuyahoga River had caught fire; deadly smog had inundated New York City; and an oil spill had ravaged the Santa Barbara coast. Rachel Carson's *Silent Spring* had rightly captured the popular imagination (while spurring the chemical lobby to hit back with the kind of disinformation campaign that is still perverting what passes for environmental policy debate in this country). In 1970 twenty million people made themselves seen and heard on Earth Day. It was the outward demonstration of widespread public outrage that had spurred Richard Nixon to announce, "The 1970s absolutely must be the years when America pays its debt to the past by reclaiming the purity of its air, its waters, and our living environment. It is literally now or never." After the real perils penetrated the popular consciousness in the late 1960s, political momentum caught up in the early 1970s. It was a wildly generative time for lawmaking.

By the end of 1973, Congress had created the Environmental Protection Agency, reinvented the Clean Air Act, and passed the Clean Water Act, the Endangered Species Act, and, equally essential, the National Environmental Policy Act (NEPA), which requires the federal government to consider and disclose the environmental effects of its actions before taking them, while ensuring the public has a voice in the decision making. Living today, it is breathtaking to imagine so much getting done in Congress. It's worth remembering that a full-blown environmental crisis was able to focus so many minds at once.

The bedrock laws passed in the 1970s imposed strong mandates, and even more inventively, they empowered you and me to enforce those mandates, recognizing that political will and government resources will too often run short. While governments generally prefer not to be sued, any person who stands to be harmed by the government's actions can hold the vastly powerful U.S. government to account in court when it fails to live up to its responsibilities. And when polluters, including the most powerful multinational corporations, violate the law, anyone in harm's way can act as a private attorney general, stepping into the government's shoes to enforce the law, even when the government won't. While they are by no means perfect, these are laws with sharp teeth, and the teeth belong to us. At a table that is set for high-flying officials and captains of industry, we can spring like tigers.

Driving fast on a highway, I get spooked by the proximity of sudden death to life as usual. The first few years of the twenty-first century felt like that. In Montana, where I was then living, you could see the ominous signs in warming trout streams and dying whitebark pines. Farther north, Arctic sea ice and permafrost were melting, and Inuit people were becoming climate refugees. Even in the uptick of dire news reports, you could suddenly feel how fast we were speeding toward the tipping points that scientists had warned us never to reach.

By 2005, colleagues and clients around the country were noticing proposals to build new coal-burning power plants. At the height of the coal rush, there were more than 160 proposed coal plants, and if built, they would have entrenched U.S. dependence on coal for decades to come. It was so precisely, so disastrously, the wrong direction to take. Coal plants were then the largest carbon polluters in the country and, notoriously, the worst contributors to the so-called "conventional" air and water pollution that was sickening and killing millions of people every year.

The massive project of blocking new coal plants overtook many lives, including mine. My first big fight was close to home. A group of rural cooperatives was proposing to build a new coal plant, the Highwood Generating Station near Great Falls, Montana. I didn't know how we would stop it. At the time, the common sense of the country was that artificially cheap coal power was essential to U.S. prosperity, and the politics were against us. I didn't yet have any real experience with energy. My prior work had been focused mainly on public lands and endangered species. In this case, we would have to challenge an air permit in an evidentiary proceeding that required the kind of courtroom skills my mom had but I didn't. I was used to writing and arguing before judges in federal court, not deposing and cross-examining witnesses. I was filled with the kind of dread that only litigation can induce, and inexperience made it infinitely worse.

But the climate imperative hooked me. And I could never say no to my fearless client and friend Anne Hedges at the Montana Environmental Information Center. She always makes things happen. There was also a self-organized cadre of amazing volunteers, including Dr. Cheryl Reichert and Jerry Taylor, who were leading the local opposition in Great Falls, and their group, Citizens for Clean Energy, became a client too. My boss at the time, Doug Honnold, told me, "Just follow your nose," implying permission to improvise and make mistakes—I did both, gaining courage along the way from my heroic colleague and mentor, Tim Preso.

We had strategy meetings in all kinds of places that felt more personal than professional—in restaurants, at my house. Anne and I were on the phone all the time: early mornings, evenings, and weekends. We were trying to develop novel legal claims and constantly having to remind ourselves: Why not? At some point, I had to stop worrying about what someone else would do in my place and just do my best. It helped when my brilliant co-counsel, Jenny Harbine, joined our team.

We did not put on a perfect case, but it was good enough to win. We also got some lucky breaks. Among them, our opposing counsel was an ethical person who turned over a damning piece of evidence in time for us to use it. There was also an experienced trial lawyer on the board who used our written filings to question witnesses and extract the admissions we needed. In the end, the board ruled for us, and Highwood's air permit was sent back for a do-over.

I believe it was important that we were part of the same community as the regulators we were suing and the members of the review board that was hearing the case. It was an ephemeral moment of political opportunity, when the reality of climate change was breaking through to all of us. I sensed a shared weight of responsibility, both personal and professional, in the hearing room. Showing up created the possibility of a different outcome.

Meanwhile, we had gone deep to understand what the financial underpinnings of the project were. We discovered the project was predicated on federal financing from a Depression-era loan program designed to electrify the rural West. With some digging, we realized that financing coal plants was in conflict with the budget approved by then-president George W. Bush, and we filed a lawsuit challenging the loan program's failure to assess and disclose environmental impacts, as federal law requires.

The government's response was to suspend the loan program, citing uncertainties related to climate change and rising construction costs. Over $1.3 billion in slated funding for Highwood and five additional proposed coal plants evaporated, fueling the true narrative that coal plants were risky investments. Ultimately, plans for the Highwood coal plant were abandoned, and the cancellation became a widely reported proof point that coal was on the ropes—which it was, thanks to all the other people like us who were fighting other proposed coal plants.

In every successful effort to make change, there is some lucky convergence of circumstances. But in my experience, there is always one essential ingredient: scrappy people who are willing to work backward from goals that seem impossibly ambitious at the start.

The critical role of lawsuits in other struggles—for instance, for civil rights and women's rights and marriage equality—is well understood.

Not so well known is the role of lawsuits in delivering cleaner air and water, preserving and restoring ecosystems, and generally arresting the human tendency to kill the world.

By invoking the "power of law," I don't mean to conjure up a silver arrow that inevitably hits its mark. On the contrary, well-resourced, historically White groups like mine have harnessed the power of law for half a century and failed to insist, much less deliver, on the imperatives of environmental justice. But as we work to repair that irreparable damage and rise to the awesome challenges that face us now, I stand by the flawed but extraordinary laws passed in the 1970s.

Those laws connect us with real power, including the power to push a transition from fossil fuels to clean energy. With the advent of Trumpism and authoritarian ambitions, we are seeing the most ruthless attacks yet on our legal framework for environmental protection, precisely because these laws connect people with their power. It is our time to make the courts and our bedrock laws a voting issue and a political passion.

I miss my mother, and I feel her absence even as family friends suggest that she is all around. But I am surrounded by leaders who wield similar skills—of connection, of practical action, of moral courage—to achieve spectacular impact. Our climate emergency requires big thinking of all kinds, but there is no single elegant solution, no perfect blueprint for a livable future. We underestimate the power of contribution—of acting within our own sphere of influence to tackle the piece of the problem that is right in front of us.

In a few decades, if we look back from a place of relative comfort and safety, I think we will remember millions of people who saw the unprecedented danger and didn't look away, who connected with their power and used it to lead change from the ground up. I write this with real people in mind. Linda Garcia, organizing her community in Vancouver, Washington, to fend off the biggest oil export terminal ever proposed in North America. Sharon Lavigne, leading a historic effort in St. James Parish, Louisiana, to block a massive petrochemical build-out. I could write down a thousand names and still be at the beginning of the list. Behind every fight that disrupts business-as-usual is an amazingly determined, often small group of people who make their luck with smart organizing and media strategies and, indispensably, the power of law, which gives us all leverage both inside and outside the courtroom. These are the triumphs that fire imagination and political will for the systemic changes that we need.

The future is likely to demand more of us than we know how to give, and we will walk through many different doors to come together in collective action that forces an adequate response from our elected leaders. As we find our way to one another, I am making my peace with never feeling at peace, and doing everything in my power to ensure that we have strong laws and fair courts on our side.

To Be of Use

Marge Piercy

The people I love the best
jump into work head first
without dallying in the shallows
and swim off with sure strokes almost out of sight.
They seem to become natives of that element,
the black sleek heads of seals
bouncing like half-submerged balls.

I love people who harness themselves, an ox to a heavy cart,
who pull like water buffalo, with massive patience,
who strain in the mud and the muck to move things forward,
who do what has to be done, again and again.

I want to be with people who submerge
in the task, who go into the fields to harvest
and work in a row and pass the bags along,
who are not parlor generals and field deserters
but move in a common rhythm
when the food must come in or the fire be put out.

The work of the world is common as mud.
Botched, it smears the hands, crumbles to dust.
But the thing worth doing well done
has a shape that satisfies, clean and evident.
Greek amphoras for wine or oil,
Hopi vases that held corn, are put in museums
but you know they were made to be used.
The pitcher cries for water to carry
and a person for work that is real.

Beyond Coal

Mary Anne Hitt

After a long day of meetings in New York, I was on a train winding along the Potomac River home to West Virginia when I got the news: The three hundredth U.S. coal plant in ten years would be retired, shut down for good. As I read my colleague's note, a wave of emotions swept over me—gratitude for those who had worked for years to reach this milestone, relief for the local families who would breathe easier after the pollution stopped, compassion for the workers concerned about their future, and hope for our climate as another massive greenhouse gas polluter committed to closing its doors.

Dolet Hills, like many of the U.S. coal plants fueling the climate crisis, was built in the backyard of a vulnerable community and was still standing even though it was struggling to compete with cheaper clean energy. Located a few miles away from the predominantly African American community of Mansfield, Louisiana, Dolet Hills had received a D grade from the NAACP in its report *Coal Blooded*, which graded the nation's coal plants on their harm to low-income communities and communities of color. One of the plant's two main owners, Southwestern Electric Power Company (SWEPCO), had recently told regulators it intended to operate the plant until 2046. The settlement the Sierra Club reached with SWEPCO would retire the plant two decades earlier, in 2026.

SWEPCO operates in Louisiana, Arkansas, and Texas and generates more than 80 percent of its power from coal, compared with the national average of 25 percent. It is a subsidiary of American Electric Power, which ranks second in the nation for coal consumption. Over seven years of advocacy in several states, we made the case that Dolet Hills not only was a notorious polluter—the worst in Louisiana's power sector—but also was costing customers millions extra each year. We uncovered that, for years, it had been dodging market rules and conducting a sneaky utility practice of forcing its expensive coal power onto the grid, despite the availability of cheaper alternatives such as wind power. Discovering these shenanigans was infuriating—but also motivating, because we could put the information in the hands of the public and decision makers to demand change.

After the settlement was finalized, 230 coal plants remained on the U.S. power grid without retirement dates. The following day, the retirement of two more plants was announced—one in Colorado and one in New Mexico, both operated by Tri-State, another famously coal-heavy utility that has been the target of years of advocacy. As I write this, we're at 315 plants down, 215 to go.

So why did these coal-loving electric utilities, some in deeply politically conservative parts of the country, agree to shut down their plants? For the same reason as the 299 coal plants that preceded them: They were boxed into a corner by a determined movement of sophisticated, tenacious advocates that made the case before regulators, elected officials, and the public—in venues ranging from public utility commissions and corporate boardrooms to editorial pages and public hearings—that the plants were dangerous, uneconomic, and obsolete. Coal is simply untenable in the twenty-first century.

This is the work of the Beyond Coal Campaign, which I have had the honor of helping to lead for the past decade. In the United States, alongside more than three hundred partner organizations, we have blocked new construction of more than two hundred proposed coal plants, secured the retirement of more than half of existing coal plants, helped usher in the era of clean electricity, driven down greenhouse gas emissions, and saved thousands of lives.

As a result of our work, we're now getting less than a quarter of our electricity from coal in the United States—down from half of our power a decade ago—and we're on track to meet the call of the world's climate scientists to phase out coal in the developed world by 2030. There are equal amounts of coal on the power grid in the United States and Europe, and there's a sister campaign across the Atlantic called Europe Beyond Coal that's working toward the same goal. Phasing out all that coal power and replacing it with renewable energy is one of the biggest near-term steps we can take right now to address the climate crisis.

Here in the United States, by retiring these coal plants, we've also tackled one of our largest sources of dangerous air and water pollution, saving more than eight thousand lives every year and preventing more than 130,000 asthma attacks annually. All the while, our lights have stayed on and our electricity bills have remained stable. We now have a choice between electricity that's dirty and expensive and electricity that's clean and cheap, and that's a no-brainer. In a world where most so-called win-wins are actually trade-offs, this one is as real as it gets.

I say all this not to brag but to rejoice that hard work—of individuals, advocates, and coalitions—does sometimes pay off in ways that really matter. In ways that we can be proud of.

Even still, there are also real losses that we have to reckon with. As a mom, I know this work is essential to giving my daughter a fighting chance at a future that's safe from the worst of climate change. As a West Virginian, I know that making these changes at the speed and scale required is creating painful challenges for some of our neighbors—challenges that our nation and our communities need to tackle head on as we transition to a clean energy economy.

My daughter is an eleventh-generation West Virginian through my husband's side of the family, and I want her to be able to raise a twelfth generation here if she chooses to do so. The climate crisis is real and it's scary, and my desire to protect her from it is why I'm working to end our reliance on fossil fuels. But I also don't want her home state to become a ghost town filled with the tattered remains of faded glory.

When you're hiking a tall mountain, there's something known as a false summit—a place that looks like the top from below, but once you get there, you see that you still have much farther to climb. That's how this moment feels to me, like both a summit and an overlook to the next massive climb ahead.

All the progress we're making is not nearly enough. We need many more successes, at an even faster pace. We have to accelerate and replicate this success economy-wide, *this decade*. And we need to do it in a way that doesn't throw people overboard and addresses the long legacy of environmental injustice. So it's worth pausing—taking a moment to consider the hard-learned lessons of the work to rid our grid of coal power, lessons that might help guide us as we continue the ascent. Here are ten of the lessons I've learned.

Advocacy—grounded in strategy—is essential.

For ten years our Sierra Club coal-retirement projections have consistently been more aggressive and more accurate than those of Goldman Sachs, the Energy Information Administration, and many other experts. That's because we have an "advocacy dial" on our model of the nation's coal fleet. We built the model back in 2010, to sort out which of the nation's then 530 coal plants we should target first, since we didn't have the resources to take them all on at once. We included all the information we could get about every coal boiler in the nation: age, size, pollution controls, economic viability, political support, environmental justice impact, and more. The advocacy dial became the campaigner's lens through which we viewed all that information, identifying the economic and political vulnerabilities of each plant.

Coal is facing increased competition from wind and solar, with battery storage, but that doesn't mean utilities are feeling the financial pressure. That's because the rules that govern utilities insulate them from market forces and give them a guaranteed profit—many enjoy a cushy monopoly in their markets—so that even uneconomic plants have little incentive to retire. It requires intentional advocacy to force utilities to act on those market signals at all, much less in the time frame that science demands. We've spent a decade learning how to pull those levers.

One of the critical truths I've learned: Market forces alone simply won't retire coal plants. "Markets will take care of it" is a persistent but incorrect claim. It's especially insidious because it strips agency from regular Americans who have figured out how to transform our electricity sector and instead insists the fate of our climate is at the whim of mysterious "market forces" that feel even further beyond our control than the vagaries of Congress.

Yes, market forces and politics absolutely matter, but smart advocacy can deliver even when they are failing us. That's exactly what we've been doing for the past decade, and what we must continue to do. The fossil fuel industry is made up of powerful incumbent players that are not giving up their market share without a fight, and we can't count on market forces to do that work for us. We actually hold the reins of power—we just need to use them.

Advocates need to understand economics.

While market forces have not been the sole driver of our shift away from coal, we've worked hard to understand them and leverage them. The two core strategies of our campaign are economic. The first hinges on a basic truth: Coal polluters pollute—coal ash, mercury, arsenic, lead, sulfur dioxide, and more—and traditionally they've dumped most of that pollution on the American public for free. Economists call these costs to society "externalities," and for decades loopholes in our clean air and water rules meant regular Americans covered the unpaid price for that pollution, sometimes with their health and even with their lives.

During the Obama years, advocacy, public pressure, and litigation closed or tightened many of those loopholes, and that forced coal polluters to pay to clean up their messes. Enforcing those standards gave coal utilities two choices: clean up a plant or retire it. And retirement is often the only economically viable choice for an aging coal plant. When coal has to clean up its pollution, it can't compete.

The second strategy forces a head-to-head competition between coal and clean energy. Utilities regularly hide economic information from the public and regulators or enjoy market rules that shield them from competition. So it's up to advocates to uncover that information, make a compelling case to regulators and the public that each individual coal plant is more expensive than other alternatives (i.e., new renewable energy), and force decision makers to act on that information.

Here's one example. The utility PacifiCorp, part of Warren Buffett's Berkshire Hathaway empire, is one of the biggest contributors to climate change in the western United States. For years it tried to hide economic data about its coal plants. Even when Oregon regulators forced it to hand over that information, PacifiCorp kept it hidden from the public. We went to court, and we commissioned our own study that confirmed what we expected—most of PacifiCorp's coal plants were indeed more expensive than wind or solar power and were costing its customers hundreds of millions of dollars more than clean energy. After years of advocacy, PacifiCorp finally relented and released its numbers, followed by an energy plan that included retirement of some of its most costly and polluting coal plants.

Beyond our power grid, as we tackle emissions across other sectors of the economy, such as transportation and buildings, finding and using these economic leverage points will be key to breaking the grip of the coal, oil, and gas industries. Given the rapid pace of innovation and the plummeting price of clean alternatives, much of the fossil fuel industry is skating on surprisingly thin economic ice. That flimsy footing provides ample opportunities for advocates, if we have the means, the know-how, and the wherewithal to seize them.

State and local decisions move the needle.

Federal policy and leadership are important, of course. But many of the big decisions that matter to our climate are made at the state and local levels. State utility commissions govern electricity production. State legislatures set clean-energy targets. Cities determine transportation policy and set building codes. State and local governments are in charge of enforcement and zoning for mining, drilling, and fracking (a destructive drilling technique that injects pressurized liquids underground to fracture rock formations and extract oil and gas).

One example comes from northern Indiana, a politically conservative state where the coal industry remains powerful and where the Northern Indiana Public Service Company (NIPSCO) was the target of years of pressure by clean energy advocates and community leaders. In 2018 NIPSCO completed an energy plan that found that by retiring its two large, polluting coal plants and replacing them with a combination of wind, solar with storage, and efficiency measures, it would save its customers more than $4 billion compared with the cost of just running the existing coal plants. NIPSCO has solicited bids for solar equal to the power generated by its coal plants, and state and local decision makers will be key in the implementation of the plan.

Focusing on those more localized decisions is how we've continued to retire coal plants and expand clean energy under the Trump administration, despite its false promises of reviving the coal industry.

But there's lots more work to do. For every decision advocates can challenge, many more pass us by because we don't yet have the resources—in terms of dollars or people power.

Local pollution motivates action.

Yes, fossil fuels are cooking our planet, but they're also creating scores of scary local pollution problems that have people up in arms: coal-ash spills in our rivers, oil slicks on our beaches, smog triggering asthma attacks among our kids and heart attacks and strokes in adults, fracking ruining our drinking water, toxic mercury in our seafood. The side effects of fossil-fueled electricity are actually front and center in our day-to-day lives.

We should continue the good work of expanding public support for climate action, but most Americans don't need a lot of convincing to phase out fossil fuels. Once they realize that pollution has landed on their doorstep and poses direct and immediate threats to their families, and that there is a cleaner option, the choice becomes obvious. And once people are engaged, it doesn't take long for them to connect the dots from neighborhoods to the atmosphere, building and strengthening the movement we need.

Here's a good example. A decade ago, there were two coal plants operating in downtown Chicago, the Fisk and Crawford plants, that were ancient, polluting, and making local residents sick in the largely Latinx neighborhoods where the plants were located, in addition to being big contributors to climate change. They were so old that Thomas Edison and Queen Mary had signed the guest book at the Fisk plant. After years of organizing and advocacy by a coalition of more than fifty organizations, including neighborhood environmental justice groups and the Sierra Club, the local elected officials who had been shielding the plant from accountability relented. The utility was forced to sit down with the city's mayor, and the plants stopped operating in 2012. Local pollution and public health concerns drove the campaign, but the outcome was a big win for the climate as well. The Fisk and Crawford were plants numbers ninety-nine and one hundred to announce retirement in the United States.

Bold and clear goals enable open-source campaigns.

I've often thought of the Beyond Coal Campaign as open-source, "something people can modify and share because its design is publicly accessible." When we started the campaign, we had a very convoluted description of our goals that was driven in part by the perceived politics of the time, when the coal industry was calling the shots in Washington. We said we were not opposed to coal as long as it was mined responsibly, burned cleanly, and disposed of properly—something that was clearly never going to happen, given the industry's century-long track record of harming workers, communities, and the environment.

In 2010 we changed the campaign name to Beyond Coal and declared our clear intention of phasing out all coal by 2030. Our audacity was breathtaking at the time, and I was practically laughed out of the room by more than one utility executive and congressional staffer. But the bold, clear goal opened up the campaign—anyone working to hold the industry accountable across its life cycle was contributing to our shared progress.

Fast forward a decade, and there are more than three hundred U.S. organizations, from neighborhood groups to large national nonprofits, that are part of the Beyond Coal network and are working toward this overarching yet very specific goal. I think of it as a diverse ecosystem where each organization plays a role, just as different organisms do in the natural world. The network includes lawyers, environmental justice leaders, economists, grassroots organizers, doctors, media strategists, data analysts, community volunteers, and campaigners. Each is essential to our progress.

Environmental justice must be central.

Communities of color and low-income communities are disproportionately harmed by fossil fuel pollution, but their voices and leadership have been marginalized by both decision makers and large environmental organizations like mine. The Sierra Club is now on a journey to learn from that history and become a better partner, and the Jemez Principles for Democratic Organizing provide a framework for thinking about our advocacy in a different way. We are not perfect, but we are committed to the path.

Environmental justice organizations are often not only the first to raise the alarm about big polluters but also instrumental in holding them accountable. I'll never forget my trips to River Rouge, Michigan, on the outskirts of Detroit, where two aging coal plants next to neighborhoods and playgrounds were operating without modern air pollution controls known as "scrubbers" and making people sick in this predominantly African American community. My heart was broken by stories of mothers who lost their children to asthma attacks, schoolkids leaving soccer fields in ambulances, and regulators who had failed these families for decades. After five years of advocacy led by local residents, the utility announced it would close the two coal plants, which had been major contributors to the area's sky-high rates of asthma.

In addition to supporting large organizations like the Sierra Club, funders need to provide sustained support to environmental justice groups, the youth climate movement, and campaigns led by those on the front lines of fossil fuel pollution. This work simply won't get done without their leadership.

We can't leave people behind.

Many parts of the country, like my home state of West Virginia, have long relied on fossil fuels for jobs and local economies. It's essential that our climate solutions provide a fair transition that supports workers and diversifies their local economies, rather than allowing fossil fuel companies to pocket big profits, declare bankruptcy, and walk away, abandoning the families and communities that sacrificed so much to power our nation. We must also provide relief for the communities of color and low-income areas that have long borne the brunt of fossil fuel pollution.

If we leave folks behind, our progress will be fragile and flawed. But if we bring everyone along, we're building a better future—one that the coal, oil, and gas industries won't be able to unravel. As a

West Virginian, I know that this transition won't be easy for everyone, and I don't want to sugarcoat the challenge we face in diversifying the economies of regions that have long depended on fossil fuels. But we're beginning to see glimmers of what's possible, and decarbonizing our economy has the potential to put millions of Americans to work.

Already more than three million Americans work in the clean-energy industry, outnumbering fossil fuel workers three to one. There is so much more work to do: retrofit our buildings, create transportation alternatives, expand regenerative agriculture, and restore the land and water that have been ravaged by mining, fracking, and drilling. Here in Appalachia, where the devastating practice of blowing up mountains for coal still continues, restoring the landscapes ravaged by "mountaintop removal" will take many years of work—and the same is true across all our coal-mining regions.

But the job of designing a fair transition is too big for philanthropy and civil society. While state and local decisions drive most of our energy choices, making the transition to clean energy a fair one requires leadership and resources at the level only the federal government can provide—leadership that is long overdue here in the world's wealthiest nation.

Winning on electricity is foundational and catalytic.

To stop runaway climate change, we have to electrify everything—cars, homes, industries—and power them with 100 percent clean energy. Electrification is the way to heat homes without fracked gas, fuel cars without gasoline, and manufacture critical goods without burning coal. That means cleaning up the power grid is essential not only to tackle a major source of climate pollution—electricity production—but also to provide the foundation for progress in other key sectors. That doesn't mean we wait to work in those areas. It means it's mission critical that we get the electric sector done right, and quickly: 100 percent clean electricity in the United States by 2030.

This has implications far beyond our borders. Many other countries around the world are not on track to meet climate targets. We sometimes like to point fingers—*What about China? What about India?*—but we're actually all in the same boat: falling short. Bold U.S. leadership is the best contribution we can make to shifting that dynamic. The world's climate scientists are clear that it's essential for the developed nations of the world to phase out coal by 2030 to keep global temperatures below critical thresholds. So far, we are on track to do that here in the United States. If we get our own house in order, innovate affordable solutions, and demonstrate that people can enjoy a high quality of life in a decarbonized economy, that will accomplish far more than endless hand-wringing about China and India. Leadership is best shown through action.

Fracked gas is a bridge to nowhere.

For years, some in the climate movement and my own organization believed gas was a necessary "bridge" to the clean energy future. But by 2012 the Sierra Club had responded to the latest science and grassroots activism to take a firm position opposing it. Today we're working with many partners to fight new gas infrastructure from the Pacific Northwest to Florida, including export facilities, new gas pipelines, and gas power plants. And in many states we're starting the work of unwinding and replacing existing gas—from power plants to home heating and appliances—with clean energy.

That's because we now know that fracked gas is not necessary and it's not useful. Our climate can't afford to build out new infrastructure based on fracked gas, especially given that it releases large amounts of methane, a "fugitive emission" that, in the twenty years after it's released, is some eighty times more heat-trapping than carbon dioxide. What's more, the local impacts of fracking, such as contaminated drinking water, toxic air pollution, and the destruction of farms and forests, have been catastrophic.

We need to phase out fracked gas as quickly as we can—and that work is actually under way. U.S. domestic gas consumption splits roughly into thirds for electricity production, buildings (mainly space and water heating), and industrial uses, such as manufacturing steel and concrete. Advocates are tackling all three, along with the 10 percent of U.S. gas that is exported.

We're applying our experience in the electric sector to do for gas what we did for coal—leverage its precarious economics and the public demand for clean energy to stop the gas rush. Advocates have begun successfully pushing cities to electrify new construction, a first important step to backing gas out of our buildings; gas has been linked to higher rates of asthma when used for cooking and heating. And we still have a window of time to block the industry's nascent efforts to create new markets in plastics (which are made from fossil fuels) and exports (to other gas-hungry countries) before they get a strong foothold.

Substantial resources deliver substantial results.

Most of the more than three hundred coal plant retirements we've secured have taken five to ten years of steady campaign work at the state, local, and regional levels. In the venues where decisions about our electricity are made, such as state utility commissions and power company planning, reaching a decision about the fate of a coal plant or renewable energy project can take months, sometimes years. This in turn requires years of focus and hard work by advocates that includes turning people out to public hearings, informing communities about what's going on and what's at stake in these often byzantine proceedings, providing expert arguments to regulators, and getting our message out in the press. It's a lot of work, but it delivers results.

In addition to that state and local work, federal advocacy is another critical long game that requires years of sustained focus and support. It involves both pushing to make sure clean air and water protections are strong and enforced by agencies like the Environmental Protection Agency (EPA) and defending them when they're under attack.

Large, multiyear grants from donors have been essential to our ability to deliver victories at this scale and pace. The next decade will see thousands of decisions that will really matter for our climate. If philanthropists invest in the resources needed for advocates—from lawyers and experts to organizers and campaigners—to influence those decisions across all sectors of the economy, our progress will be exponential.

Finally, if there's one overarching message I want to share, this is it: *We can do this, and it's not too late.*

In just a decade, our grassroots movement did the impossible and moved one of the most powerful fossil fuel industries in the world toward obsolescence, despite a hostile political and economic landscape, at a time when the warming of our planet was a low priority for most Americans. Today the winds of innovation, economics, and public will are at our back, and the climate crisis is impossible to ignore.

We are the architects of our future—not the fossil fuel industry. I say that not out of sentimental aspiration but because of what we've already done and how much more I know we're capable of accomplishing. Momentum is on our side. Transformation that can avert the climate crisis is possible in the decade ahead. Let's go make it happen.

Whether we suffer greatly or build together—it is our choice. . . . We must choose whether this moment will lead us to regression or evolution, authoritarianism or greater democracy, extraction or preservation. Our greatest choice is to move towards a cooperative, collaborative world that aligns with scientific consensus.

—Rep. Alexandria Ocasio-Cortez

Collards Are Just as Good as Kale

HEATHER McTEER TONEY

Many people think of "environmentalists" as White people hugging trees, and of "the environment" as the forest or jungle as opposed to their own backyard. My relationship with the natural world, my people's relationship, is a swirl of gratitude, trauma, and spiritual connection. My Black, southern, rural ancestors connected to land and soil in ways that are both good and bad but almost always, and most of all, powerful.

Black folks have always had a deep and physical connection to the environment. The land that our ancestors were forced to work was the very same space where they lived. The field where our mothers toiled was often the place where they also gave birth. Our history has entwined us with the land in a profound way, and our connection to the land is as symbiotic as bees to flowers. Yet our voices are constantly ignored on matters concerning climate impacts and environmental protections.

Mainstream America ignores hard-learned lessons from a people who, enslaved and forced to travel across the Atlantic under unimaginable conditions, to work a land they did not know, figured out how to make things thrive and grow using their wisdom of nature and spirit. As we all face a climate crisis that threatens our very existence, the ability of my ancestors to adapt to wholly new environments offers wisdom to embrace.

I'm from an agrarian part of the country. The Mississippi Delta is home to some of the world's most fertile soil. My hometown, Greenville, Mississippi, sits midway up the state, bordered on one side by the powerful Mississippi River. Growing up, nature was a part of everything we did, whether we knew it or not. I knew when deer season started and ended, not because my family hunted but because our neighbors would take weeks to prepare and ready themselves for opening day. I could tell when the cotton and soybeans had been planted, not because I was a farmer but because of the constant roar of crop dusters overhead, spraying pesticides from small planes. I knew that dark leafy greens came during the fall, squash in the spring, and tomatoes in summer—not because I studied horticulture but because my godmother, Mrs. Loubirtha Irvin, kept a garden visible from her kitchen windows. I chuckle thinking of American society's recent lovefest with kale—and how people are still in the dark about collard greens, which have been around just as long and are just as good (better, if you ask me). From homegrown vegetables and pecan picking in the fall to knowing full well that mild winters meant the mosquitoes would be relentless in the summer, everywhere I turned I was surrounded by the interweaving of nature with Black culture, poverty, and the rural South.

It is the Black part of American culture that is especially precious to me but largely missing from the public conversation around environment and climate solutions. Environmental justice is a fundamental civil rights issue, and the rural South is plagued with so many critical needs and civil rights concerns that it's hard to keep up.

In 1969 my father came to Mississippi straight out of law school to help establish voting rights for people of the Delta, and my mother accompanied him to teach school. The conditions they embraced upon their arrival still exist today. In addition to the extreme poverty that has plagued the Delta for generations, the rights of poor people were being trampled—voter intimidation, racial discrimination, and educational disparities. People wanted to believe the promise of jobs and economic security peddled by big companies, but their operations yielded polluted land and water. The land became increasingly toxic; food insecurity was rife. In some ways, little has changed in the intervening decades. Our local elected leaders remain overwhelmed and underfunded as they face the struggles of maintaining basic infrastructure like streets and sewers, all while living with the knowledge that they need to somehow build resilience in order to survive worsening storms and other climate impacts.

When I look back at my evolution within environmental work, it has emerged from understanding that all of this—ancestry, nature, faith, and civil rights—is about community. Community is the lens so often left out of the environmental discussion, but it's vital for identifying real solutions.

My community spoke the language of faith. That's what kept us believing that things would turn around someday. Whether it be Jesus, a job, or a young Black girl known for being smart, adventurous, and a bit sassy, empowerment would come. You will not find a place filled with more love, resilience, and faith than Black communities in the Delta.

The same faith that was evident in my childhood continues to be relevant in the work of climate and, particularly, climate justice. Of all the scripture I heard growing up, the one passage I will never forget comes from the book of Hebrews:

> Now faith is assurance of things hoped for, a conviction of things not seen. . . . By faith we understand that the worlds have been framed by the word of God, so that what is seen hath not been made out of things which appear.

We repeated these words almost every Sunday at Agape Storge Christian Center, the nondenominational church where I was raised. There I learned not only of God's grace and mercy but also of his expectation of humanity to love all of creation and to care for it as God cares for us. As a Christian, I was taught that this was the only way to enact my faith, and with faith, nothing is impossible.

But there is an important accompaniment to this scripture that we often forget:

> For as the body apart from the spirit is dead, even so faith apart from works is dead.

In other words, as our pastor's wife would often lovingly say:

> You can pray and believe all you want, but without action, "ain't nothing about to happen." You just wasting the Lord's sweet, precious time.

We can plant seeds, but without watering and tending the garden, nothing grows. We can talk about how awful our elected officials are on issues of climate justice and civil rights, but if we don't go vote . . .

So where was the faith for our environment? Unbeknownst to

some, you can believe in Jesus and accept the reality of climate science at the same time. Throughout my journey, my faith has made it more and more evident that it is my responsibility as a Christian to take care of what God has blessed me with, including my place on this Earth. We've done a disservice to the environmental movement by couching Christianity as dominionism (the idea that God has given us dominion or rulership over the Earth) as opposed to creation care (the understanding that God has charged us with the care and stewardship of creation, and that its prosperity is tied to our own). This is often a point of difference between conservative evangelical (and typically White) Christian theology and a more progressive understanding of the Bible often found in the Black church. The tenets of my faith require that we "tend and keep" God's creation. The idea of salvation is an idea of freedom that carries with it the sense of responsibility to care for all things, to care for one another better.

During my time as mayor of Greenville, Mississippi, 2004–2012, I had to display and use that faith more than ever. I grew from seeing nature and the environment as just part of life to respecting my responsibility to care for and protect it. While well aware of the environmental injustices that have plagued Black communities for years, I did not foresee the dramatic climate impacts my state would face during my term. Lisa Jackson, then administrator of the U.S. Environmental Protection Agency (EPA), visited my community and later asked me to serve as chairwoman for the EPA's Local Government Advisory Committee.

Two weeks after I accepted the appointment, the BP oil spill occurred. Talk about a trial by fire—I was charged with convening local elected officials across the Gulf Coast and providing advice to the EPA on coastal cleanup from a local perspective. All this as southern cities and towns, not quite ready to accept the reality of climate change, were increasingly experiencing its effects. Over the course of my eight years as mayor, Greenville also experienced two historic flood events. Not since the great flood of 1927, when the levees broke and both lives and land were lost, had our town seen such water. Now, though, this has become a relatively common occurrence. We weren't waiting for climate change; climate change had come.

Despite hearing the Republican rhetoric of "climate change ain't real," people knew that something more than a rising river was changing and amiss. The river waters were coming faster and stronger from the increased snow up north. (Heavier wintertime precipitation is yet another outcome of rising global temperatures.) Each time Chicago, Minneapolis, and other midwestern cities got strong winter storms, the snow melted into streams that eventually made their way to the Mississippi Delta. Deer and duck seasons weren't the same as in years past. Cotton and soybean crop yields were different. Increased heat, droughts, and floods meant more pests. Meanwhile, it felt like no one was listening to the voices of the poor, of rural folks, of southerners.

I was later selected to lead the EPA's Southeast Region, which includes eight states and six federally recognized tribes. My new boss, then administrator Gina McCarthy, made it clear: She didn't need shy and demure. She needed leadership that had local people in mind and could listen with an ear to help. We immediately got to work meeting with mayors and community leaders and addressing the region's toughest issues. It wasn't easy, but I had pulled together a strong team. My immediate office consisted of four attorneys, two financial specialists, three community leaders, and experts in everything from Indigenous communities to particulate matter in the air.

This team was also 90 percent Black women. All too often I secretly wished for a hidden camera to take a picture of the faces around our conference room table when a guest realized that *we* were the leadership team. And while our team was not perfect, we agreed that elevating and empowering the voices of the community was our priority. We would create seats at the table—even give up our own if need be. The issues around environmental justice and climate solutions would take each one of us, and there was no room for petty division.

At the end of 2015, I became pregnant with my first child. I was excited and fearful of what it meant. Children are especially susceptible to things like heat, asthma, allergies, and insect-borne diseases, which are all made worse by extreme weather and emissions. Behavioral and mental health challenges have also been directly linked to a worsening climate. Studies even connect climate change to violent crime. Today's evidence of hot weather being linked to increased shootings is a tragic "I told you so."

Being pregnant brought me closer to other issues too. The idea that my unborn son could suffer harm from something like the Zika virus, a disease made worse by climate change, made every insect an enemy. I listened with new ears when talking with mothers who were farm workers in Florida. Their fear of working around pesticides, and the pains they took to make sure they didn't bring home harmful chemicals on their clothes, took on new meaning. My heart ached for the mothers of Flint, Michigan, dealing with an ongoing water crisis. My environmental work was no longer simply to develop and implement sound science policy and good governance—it was to save lives and ensure our planet is habitable for kids in the years to come.

My fear became my focus. After leaving the EPA at the end of the Obama administration, I joined Moms Clean Air Force. We engage parents in work to move cities and towns to create municipal climate action plans and build support for clean energy and green spaces. Our community of more than 1.2 million moms and dads across the country is dedicated to reducing air pollution and protecting all children from the climate crisis we're facing. We are demanding 100 percent clean energy and no additional climate pollution in the United States as soon as possible. We are combating the draconian funding cuts and regulatory weakening that harm children's health. We protested right alongside our own children as they marched for climate action in strikes across America on September 20, 2019. Through every effort and every channel, we are encouraging more and more parents to get involved.

My faith keeps me focused. And through it all, I know: Climate change is a threat to Black life.

African Americans, Black women—particularly southern Black women—are no strangers to environmental activism. Don't believe the contrary stereotypes. Many of us live in communities with polluted air and water. Many work in industries from housekeeping to hairdressing where we are surrounded by toxic chemicals. We live in food deserts with limited healthy food options, and the options we do have are often laden with pesticides and growth hormones not intended for our well-being.

Caring about climate change is not a bougie Black thing. Who exactly do you think is on the front lines every single day, fighting to keep our communities safe from industries, polluters, and those seeking to harm our kids? Who do you think is fighting that filthy incinerator at the end of the neighborhood block, that coal-ash pile on the way to work, or that cement plant next door to the church? Who do you think has to take care of children who are made sick on their own playgrounds but can't afford to rush to emergency rooms?

We live in pollution, play around it, work for it, and pray against it. Hell, we even sing about it. Black women are everyday environmentalists; we are climate leaders. We just don't get the headlines too often.

Rarely do we see or hear Black voices as part of national conversations about climate policy, the green economy, or clean energy—even though 57 percent of Black Americans are concerned or alarmed about climate change, compared to 49 percent of White folks. We're relegated to providing an official comment on environmental justice issues like the water crisis in Flint, or we're the faces in the photos when candidates need to show that they're inclusive. Fortunately, this is slowly changing as more and more women of color step loudly up to the table and make their expertise known in climate justice and culturally competent, solution-based thinking.

For decades Dr. Beverly Wright, a professor of sociology, has been training leaders from our country's historically Black colleges and universities (HBCUs) through the Deep South Center for Environmental Justice. Her students have assisted Hurricane Katrina victims, researched climate impacts on vulnerable communities, and taken their brilliance to places like the United Nations climate negotiations in Paris. Dr. Wright is truly a sage teacher of the environmental justice movement.

Catherine Garcia Flowers fled her native New Orleans during Hurricane Katrina, only to arrive in Houston, Texas, and get bombarded by a storm once more. Instead of moving again, she became involved in climate advocacy. Catherine ran for city council in 2019 and, though she lost, her campaign elevated the issues of climate change among people in her district. A first-generation woman of both African American and Honduran heritage, she works tirelessly as the Houston-area team lead for Moms Clean Air Force and reminds me constantly that "common sense isn't common." She sees her role as "creative problem solver" and helps leaders realize that sometimes cultural solutions are found simply by validating the trauma and feelings of the underserved and harmed.

Dr. Mildred McClain is whom I picture when I imagine a matriarch of the environmental justice movement. Arrayed in colorful skirts and beautiful wraps, her presence speaks of strength, struggle, and perseverance. It's for good reason: For three decades Dr. McClain has been the voice of resilience throughout the Savannah, Georgia, community. When the air was thick with pollution from the shipping channels in the Savannah port, Dr. McClain convened community meetings so that people were part of the solution. She encouraged African Americans in her community to become certified in environmental fields like hazardous-waste removal, soil remediation, and air monitoring. Not only is she an environmental justice leader, but she is our environmental justice griot, the holder of stories past and instructor of how to survive what is to come.

In 2019 I attended the Woman's Auxiliary meeting of the National Baptist Convention in Jackson, Mississippi. A group of us from Moms Clean Air Force had produced a climate-focused Bible study, including scriptures, lessons, and ways to activate communities on climate. Each lesson offered actions that members could take either individually or collectively as a congregation—from simple nature walks with a Sunday-school class to calculating your ecological footprint to discussing air pollution and asthma.

The content was on point, but we'd made a miscalculation. Over 350 women leaders of Black churches showed up for our session. We quickly ran out of the fifty Bible study books we had on hand.

"Look, baby, stick this in your bra." I looked over as a stout lady with a serious-looking smile leaned over to my colleague and handed her a small piece of folded paper with her name and address written inside. We laughed, as we knew exactly what she meant.

Only items of precious value and importance are kept in a Black woman's bra. Be it a twenty-dollar bill, the phone number of a special person, or numbers to be played in the lottery, there is no more secure place on this planet. To be given a note to put in one's bra, close to the heart, is an unspoken message of trust, and this dear lady was communicating it clearly. My friend stuck the paper in her bosom. She'd make sure the woman got her book.

Across the country, women of all colors and creeds are not waiting for anyone to tell us what we should do about the climate crisis. We're giving directions, demanding action, and riding into battle to save our children and our country from the impending destruction we know may result if there is not a drastic reduction in greenhouse gas emissions and a cessation of global temperature rise. We know we don't have time to sit around and wait for someone else to make decisions. Instead, we are simply finding ways in our communities to effect change. We are sitting on sustainability boards and commissions, studying the science, galvanizing our friends and neighbors, and teaching Sunday-school classes. We're finding every foothold of action available because our communities, our mothers, and our daughters are depending on us, and because our ancestors are watching. We shall make them all proud.

For Those Who Would Govern

JOY HARJO

First question: Can you first govern yourself?

Second question: What is the state of your own household?

Third question: Do you have a proven record of community service and compassionate acts?

Fourth question: Do you know the history and laws of your principalities?

Fifth question: Do you follow sound principles? Look for fresh vision to lift all the inhabitants of the land, including animals, plants, elements, all who share this earth?

Sixth question: Are you owned by lawyers, bankers, insurance agents, lobbyists, or other politicians, anyone else who would unfairly profit by your decisions?

Seventh question: Do you have authority by the original keepers of the lands, those who obey natural law and are in the service of the lands on which you stand?

The Politics of Policy

Maggie Thomas

When the Warren campaign brought me on as climate policy adviser, Theresa Landrum was the first call I made. Senator Elizabeth Warren would be coming to Detroit. Would Theresa be interested in meeting the senator and leading a walking tour of her neighborhood? Senator Warren and her staff were in the process of writing an environmental justice plan and wanted to learn directly from frontline communities, like this one in the shadow of a Marathon oil refinery, about how the federal government could meaningfully address the long legacy of environmental racism.

Months earlier, when I was one of the climate advisers to Governor Jay Inslee's presidential campaign, it was Theresa and her fight for environmental justice in Detroit's 48217 neighborhood—called by its zip code, the most polluted in Michigan—who taught me that good climate policy must be rooted in a culture of listening. We met in the Detroit summer heat ahead of Governor Inslee's visit. Theresa walked me along the fence line of the refinery, which takes center stage in this community. We stopped at the top of a freeway overpass. As far as the eye could see, Detroit—Motor City—was peppered with smokestacks and petrochemical plants, all puffing along.

Theresa and I talked for well over an hour. She knew every twist and turn of local permitting, chemical plant expansions, real estate buyouts that fall along racial lines, and the myriad injustices her community faces. She told me stories of cancer survivors, herself included, and loss. Childhood asthma, adult asthma. Clean air is systemically out of reach in Theresa's predominantly Black neighborhood. It was obvious—this community was being poisoned. And her ask was clear: Listen to the communities most affected by environmental impacts when crafting policy, because nobody knows better the nuances of our struggles, or the solutions that will lead to a more equitable future, than those affected.

It will never be enough to include environmental justice as an afterthought. Despite government attempts to decrease pollution, studies continue to show Black families are more likely than White families to live in neighborhoods with high concentrations of air pollution—even when they have the same or more income. There is more than enough data to show that environmental injustices plague communities across America. But listening to Theresa showed me the extent to which environmental racism is at the root of so many other issues—housing, poverty, education, health—and that solutions to our country's most wicked problems must recognize how these issues are interrelated.

By listening to Theresa and many environmental justice leaders from around the country, we crafted Governor Inslee's "Community Climate Justice" plan to focus on how communities most affected by climate change and industrial pollution can lead in the federal government's effort to address America's legacy of environmental racism and historic disinvestment. After many similar conversations, we wrote Senator Warren's plan for "Fighting for Justice as We Combat the Climate Crisis" to give advocates like Theresa a direct line to the White House through a National Environmental Justice Advisory Council, hardwiring listening into policy development and implementation. This plan would enable frontline community leaders to offer proven solutions directly to the highest levels of government, and if policies do not have the intended effect, the council is designed to hold the sitting administration accountable. Concrete policy proposals can and should evolve out of conversations with community leaders, starting with their concerns and priorities and including solutions that are formulated together with those who are most impacted.

Governor Inslee focused his candidacy on the climate crisis—a first for a presidential campaign—putting forward a vision for a 100 percent clean energy future for America. He injected into national debates and press coverage the reality that the climate crisis is intimately related to so many of America's struggles. But campaign climate plans and policy proposals cannot just be about renewable-energy technologies and electric cars or government subsidies and taxes; they need to be about people.

When Governor Inslee left the race in August 2019, Senator Warren quickly adopted his "100% Clean Energy for America" plan, carrying his vision forward and building on it with fourteen climate plans of her own. We can't allow climate policy to be about ego and credit; it must be about collaborating, building on one another's work, and elevating the best ideas we have to solve the biggest problem humanity has ever faced. The loudest voices, those whose profits and traditional business models are put most at risk by a transition to clean energy, are unlikely to demand the transformational policies required to rebuild a just and equitable clean-energy future. In fact, it may be the unheard voices—voices that have been suppressed for generations—who have insight into critical solutions.

After Senator Warren released her plan for "A New Farm Economy," more than sixty Black farmers, advocates, and academics sent a letter to the campaign rightly identifying that, while the plan laid out some policy solutions specifically for Black farmers, it failed to address the unique legacy of structural racism embedded in both our agricultural economy and in the U.S. Department of Agriculture (USDA). The plan intended to incentivize farmers to play a larger role in our fight against climate change and break big agribusinesses' stranglehold, which squeezes farmers and farmworkers and jeopardizes the productivity of the land. But building an inclusive new farm economy requires understanding why and how the structures of the current farm economy have shut people out. Due to discrimination, especially from the USDA, Black farmers have lost 80 percent of their farmland over the last century, amounting to around thirty million acres and hundreds of billions of dollars in lost wealth. The campaign took the criticism to heart and fielded calls from Black farmers who had experienced discrimination, advocates who had been fighting for their rights, and academics who had documented the century-long discrimination. The update to the plan—"Addressing Discrimination and Ensuring Equity for Farmers of Color"—rolled out three months after the campaign received the letter.

Campaigns often avoid updating their policy positions for fear of being accused of opportunism or, worse, vacillating. Yet this kind of iteration is precisely what builds stronger, more effective, and more equitable policy. Updating this plan created an opening to shine a light on deep, structural racism and put forward bold ideas for the nation to consider. Listening to those most affected by a legacy of discriminatory policies meant we were able to put forward solutions that Black farmers on the front lines had long been advocating for—such as radically restructuring the office within the USDA that handles civil rights to demand justice for farmers of color, and setting up a federally backed land trust to buy land from retiring farmers and then sell it interest-free to farmers of color. Most important, this act of listening, of hearing, is an exercise in humility—something that politics could use a lot more of.

Listening also fueled the creation of the Blue New Deal. During CNN's climate-crisis town hall in September 2019, regenerative ocean farmer Bren Smith raised the point that the ocean was all but left out of the Green New Deal resolution—it gets a single passing mention. Bren asked Senator Warren if she would support a Blue New Deal, given the oceans' importance to both climatic systems and our economy. She immediately saw the need for this, exclaiming, "He's got it exactly right, we need a Blue New Deal."

Nearly 40 percent of Americans live in coastal counties, where lives, homes, and infrastructure are at risk. Climate change is warming the waters, disrupting migration patterns of marine life, killing coral reefs, strengthening storms, and fueling sea level rise—all problems that need a policy response. And if we don't consider the ocean, we miss out on a suite of powerful climate solutions—from offshore wind energy to restoring coastal ecosystems that absorb carbon and help buffer impacts of storms to regenerative ocean farming. So the campaign took on this idea. We sought input from marine biologists, ocean policy experts, leading lawyers in renewable-energy regulation, and labor unions that represent industries most reliant on our oceans.

The Blue New Deal became one of Senator Warren's most popular plans. It was praised frequently in selfie lines, and its ideas have been broadly embraced by supporters of other candidates because it filled an important gap. Listening allowed for an important expansion of the Green New Deal. In this case, that listening was aided by a nationally televised town hall dedicated to climate—a first in a presidential election cycle, thanks to organizing and activism across the climate community.

If we dare to listen, we can embrace climate policy as a living document—an evolving, improving set of ideas. If our planet is built upon fluid systems and cycles, why shouldn't the policies we put in place to protect it be the same?

It was the honor of a lifetime to be a part of two 2020 presidential campaign teams that fleshed out ambitious and pragmatic visions of what can be done with climate leadership in the White House. The work involved long hours calling organizations of all shapes, sizes, and spheres of influence to understand their righteous fights in advance of developing our written policy plans. It meant cultivating and relying on a group of experts to give critical feedback, test ideas, and, all too frequently, ask for help on impossibly short deadlines. And it meant doing my best to distill the sometimes disparate views of the climate community into practical political recommendations that always held a true north to do right by our planet. Although these presidential campaigns failed in the traditional sense—neither candidate secured the Democratic nomination—they increased the scope and nuance of the public discourse on climate and built detailed policy proposals that have set a new bar for our nation.

Admittedly, the old bar was quite low. In the entire 2016 election cycle, climate change and environmental issues got only five minutes and twenty-seven seconds of discussion on the presidential debate stage. But long before the 2020 election cycle began, advocates had been working tirelessly to increase electoral dialogue on the climate crisis. Politicians, including Senator Bernie Sanders, have been speaking out on the issue for years, if not decades; organizations like NextGen America, the Sierra Club, the League of Conservation Voters, and the Sunrise Movement have galvanized public support to a point where politicians know that they can't ignore climate action.

It's worth taking a moment to appreciate that 2020 may be an inflection point for climate in politics. This is the first Democratic primary in our nation's history in which nearly every major candidate had a climate plan. And this is the first presidential election in which climate has become a voting priority, regularly ranking among the most important issues for Democratic voters. For many young voters, whose futures are linked with the fate of the planet, voting for a presidential candidate with the strongest climate plan is paramount. Although scientists have been telling us for decades that greenhouse gases are causing irreversible damage to our Earth, the issue finally took the national spotlight after a perfect storm of advocacy from the climate community, scientific reports, and youth-led climate strikes around the globe. And for the first time, several presidential candidates ran on platforms rooted in the belief that the climate crisis is *the* existential crisis of our time. This gives me a renewed sense of hope for what climate policy can look like in the years ahead.

A 100 percent clean-energy and climate-resilient future is possible, but only if the climate policy to get us there starts with policy makers and staffers who are willing to listen. I have had the great fortune of being the ears of two presidential campaigns. Every conversation and outreach call was a fact-finding mission about how the federal government could help communities address the climate crisis. Sometimes that means calling for the federal government to step back, sometimes for it to step up. Listening brought me to Theresa Landrum and 48217. Theresa's fight became my fight and Governor Inslee's fight and Senator Warren's fight. Finally, Americans are seeing that this is *our* fight.

Inequality and climate change are the twin challenges of our time, and more democracy is the answer to both.

—Heather McGhee

A Green New Deal for All of Us

Rhiana Gunn-Wright

People often ask me why I decided to help develop the Green New Deal (GND). Why did I, a twenty-something Black woman, think I could help develop a policy proposal to address something as big as climate change? Often I think they expect some grand story: about incredible courage and audacity or deep ambition or, sometimes, a master plan for the revolution. The truth is that I was scared—and I really needed a job.

I grew up in Englewood, on the South Side of Chicago, raised by my mother and grandmother in the same house that my mother grew up in. In the thirty years between my grandparents' moving in with their three babes and my being born, Englewood had gone from being a (mostly) middle-income community, close-knit and quiet, to one of the poorest, most barren parts of the city. My neighborhood had so many problems: poverty, unemployment, underfunded schools, police brutality, pollution, violence. And those were just the big ones.

I rarely saw anyone in power try to solve the problems in Englewood. And when they did try, it seemed to make things worse.

When I asked my mom and grandma why Englewood looked like this, they told me about the government. About how the highway system had been built through Black neighborhoods, destroying communities that would never be rebuilt. About the city's housing authority razing public housing and scattering families in the name of "urban development," only for city officials to turn around and sell the prime real estate to developers on the cheap. About the city systematically underfunding Black schools and then shutting them down because of "underperformance." And that's just what happened to my neighborhood—not even what happened to my family.

My grandmother's family was not eligible for Social Security for at least fifteen years because her mother was a washerwoman, and that New Deal program excluded agricultural and domestic workers (nearly all Black at the time) because President Roosevelt needed to secure votes from southern Democrats, and southern Democrats needed cheap labor from economically vulnerable Black people.

My grandfather bought our house without any help from the GI Bill, despite being a veteran of the Korean War. My mother told me that he was too proud to apply. The truth is, pride or not, the government systematically denied home loans to Black veterans, and the notorious redlining in Chicago meant that he wouldn't have been approved anyway.

I grew up in a frontline community—an area close to a pollution source with high levels of air pollution. I developed asthma, like most of my friends in my neighborhood. I could barely run until I was in my late teens, and I missed school every spring. My lungs are weakened to this day.

Progress came with a price, and the price was us. By the time the GND came into my life, I would be damned before I paid another dime.

I have spent my life trying to rewrite systems of power, and policy is nothing if not a system for creating and distributing power. That is why, contrary to popular belief, the most important part of a policy proposal is not the details—at least not at the beginning. It's the *vision* that the policy presents and the *story* it tells. The best policy proposals—that is, the proposals that move the most people to fight for them—present a clear narrative about what went wrong, why it went wrong, and how the government plans to fix it.

The GND is a policy proposal for a ten-year *economic mobilization* to rapidly transition the United States to a zero-carbon economy, and in doing so to regenerate and reorganize our economy in ways that significantly reduce income inequality and redress legacies of systemic oppression. The GND's five big goals are laid out in "House Resolution 109":

It is the duty of the Federal Government to create a Green New Deal—

A. to achieve net-zero greenhouse gas emissions through a fair and just transition for all communities and workers;
B. to create millions of good, high-wage jobs and ensure prosperity and economic security for all people of the United States;
C. to invest in the infrastructure and industry of the United States to sustainably meet the challenges of the 21st century;
D. to secure clean air and water, climate and community resiliency, healthy food, access to nature, and a sustainable environment;
E. to promote justice and equity by stopping current, preventing future, and repairing historic oppression of frontline and vulnerable communities, including Indigenous peoples, communities of color, migrant communities, deindustrialized communities, depopulated rural communities, the poor, low-income workers, women, the elderly, the unhoused, people with disabilities, and youth.

The GND tries to achieve these goals in two ways. The first is through a set of *projects* that, if completed, would nearly eliminate carbon emissions in the United States. The second is through a set of *policies* that aim to both reduce inequity *and* protect Americans—especially vulnerable Americans—from the disruption and instability that transitioning away from fossil fuels will create.

Some people like to refer to the first set of projects as the "green" part of the GND and the second as the "New Deal" part. While this may be a helpful rhetorical device, it is a dangerous way to conceptualize the GND. All parts of the GND contribute to decarbonization—even the more "social" policies like universal healthcare and education and training. Similarly, the "green" projects can help reduce inequity if they are designed as the GND proposes—that is, in ways that create millions of well-paying jobs, bolster worker power, invest in local communities, and strengthen the social safety net.

Our insistence that the GND address the environment, economy, and equity *at the same time* has led many people to accuse the GND of being, at best, a progressive boondoggle and, at worst, little more than an ideological pipe dream. Why does a plan about tackling the climate crisis also try to address oppression and inequality? Is this just an attempt to force through progressive priorities under the guise of climate policy? How would we even pass something like this, given the siloed nature of congressional committees and federal agencies?

These critiques make sense—*if* you define policy only as a statement of what the government should do. But public policy is much, much more than that. Policy making is not a science. It is a political process, not just a set of solutions. If politics is a fight to elect people who reflect and share our values, policy is a fight to actually enact those values—to mold the world, through the work of government, into what we think it should be. It's a fight to see whose problems get addressed, how those problems are addressed, and who benefits.

Let's use a medical metaphor: You go to the doctor with a problem, she diagnoses it, and then you two decide the best treatment. Creating policy is more like going to the doctor with a problem, having fifteen people argue about whether it's a "real" problem, then having five of those people (plus some new strangers!) start arguing anew about what the cause of the problem is, only to be interrupted by the doctor's boss coming in to tell them that they can choose only two of five possible treatment options because the other three would cost too much.

So an insistence on treating policy *only* as a set of solutions ignores both the reality of policy making and the very aspects of policy that make it such a powerful—and dangerous—tool. I did not and do not have the luxury of that kind of ignorance, not before the GND and certainly not now, as we teeter on the brink of utterly preventable catastrophe.

To understand the GND, you have to understand the *problems* it addresses, the *principles* that guide it, and how it intends to shift *power*.

Problems

The GND is designed, first and foremost, to address the climate crisis at the speed, scale, and scope required to prevent catastrophic levels of warming. That is, it is grounded in climate science. According to the 2018 Intergovernmental Panel on Climate Change special report, global emissions have to be roughly halved by 2030 to have a chance of keeping warming under 1.5 degrees Celsius.

The climate crisis is entangled with other crises, named in the GND resolution: life expectancy declining; basic needs out of the reach of many; the greatest income inequality since the 1920s; a large racial wealth divide; a gender earnings gap; and systemic injustices that have been exacerbated by pollution, environmental destruction, and climate change. The GND emerges from an analysis of the climate crisis that identifies it as a consequence of systems—neoliberalism, strategic racism, unfettered capitalism—and their interaction, rather than simply as a consequence of greenhouse gas emissions. That is why the GND attempts to respond not only to the identified crises, climate and otherwise, but also to the economic and social disruption that will inevitably result from a massive effort to transition away from fossil fuels.

Principles

Science can help us to understand the extent of the climate crisis, identify its causes, and measure its severity. It can even suggest time lines for action. But it cannot tell us what policy solutions to pursue. It cannot tell us what to do—not definitively. That is a matter of principles.

It is possible that in addressing the climate crisis we will increase inequity. In fact, that's the default outcome, given the inertia of our current systems. How we decarbonize, how we shield our communities from climate disasters, how we transition our economy—all of these things *matter*.

I always get nervous when I talk about the values at the heart of the GND. Although it is not—and has never been—true, there is still immense pressure to pretend as though public policy were "objective": a collection of facts and details and nothing more. But the GND, like most progressive policy, is not considered objective. It is "ideological": an outburst of youthful exuberance and naïveté, driven by emotion instead of analysis, strategy, and wisdom. It is, we are told, policy for people who campaign—not people who govern.

But the truth is that our values did inform the vision of the GND—especially ideals of compassion, dignity, and justice. The United States is a nation of scarcity, and increasingly so. Seventy-eight percent of Americans live paycheck to paycheck. As of 2018, about 40 percent of Americans could not afford an unexpected $400 expense without going into debt or having to sell off their possessions. About 25 percent of Americans skipped necessary medical care because they couldn't afford it. For most people in this country, we are not a nation of prosperity. We are a nation of lack, and a world where resources are increasingly constrained by the climate crisis. If we do not counter this scarcity, how will we build anything but a society of fortresses as the planet continues to warm?

Power

The vision of power at the heart of the GND is one of redistribution: from private to public, from employer to worker, from the historically advantaged to the historically disadvantaged. That's because the ability to burn fossil fuels with no limit and no legal repercussions requires two things. First, fossil fuel industries and those who control them (or profit deeply from them) can concentrate enough wealth and political power to override the will of the people—who, by and large, want to stop climate change. Second, there are people and places that can be hurt, even killed, with little consequence. In short, fossil fuel impunity requires intense concentrations of economic and political power among corporations and the wealthy who profit from them.

Over the last forty years, the top 1 percent—whom we will refer to as "the elite"—have captured "more unfettered political, cultural, and intellectual power than at any point since the 1920s," according to Naomi Klein. This has happened, at least in part, because deregulation and other neoliberal economic policies allowed elites to accrue nearly all of the economic gains since 1980. They have used those gains in turn to finance campaigns—and politicians—and to decimate organized labor. The result has been wage stagnation and rampant economic inequality, with declines in union membership alone accounting for as much as one-third of the growth in income inequality for men and one-fifth for women since 1972. The erosion of worker power particularly decimated communities of color as the 2008 housing crisis, itself a product of deregulation, wiped out generational wealth, leaving many communities of color no better off than they were prior to the civil rights movement.

The disempowerment of workers and the resulting economic instability have made it difficult to advance climate action. People who are preoccupied with trying to feed their families simply don't have time to think about much else, particularly when they are told that "saving the environment" might mean losing their jobs. And it is not just workers who are afraid of the job-killing effects of robust climate policy. State and local governments are too.

Inadequate wages have weakened local tax bases while increasing the need for services, leaving many state and local governments strapped for cash and more reliant on outside funding—making them particularly vulnerable to the influence of moneyed interests. The result has been an erosion of local and community control over crucial policy decisions. A 2019 study of one million federal and state bills found that ten thousand of these bills were nearly exact copies of model legislation, most of which was produced and lobbied for by special-interest groups and think tanks funded by elites.

The success of the GND depends on the ability to reroute power away from the 1 percent and back to the 99 percent and the political and economic institutions designed to serve them. If we are going to become an economy that serves people and the planet, then the people—*all* of the people—need power, and we need it now.

Let's be blunt: Our current economy is predicated on a reliance on fossil fuels. They are our primary energy source. Imagine if we, as humans, stopped eating food as our primary energy source and instead had to eat red algae. How would our lives change? Where would we get the algae from? Who would grow it? How would we ship it? How much would it cost? Would we even need refrigerators anymore?!

Transitioning away from fossil fuels is no different. Fossil fuels do not just power our homes and cars. They power *everything*, from manufacturing clothing to streaming Netflix. That means that transitioning away from fossil fuels will directly affect the livelihoods of workers in the most visible fossil fuel industries—like coal, oil, and gas—*and* indirectly affect nearly every other sector. Ending our use of fossil fuels will, by its very nature, cause significant economic disruption and transformation—especially now that we have so little time. The question is simply how we will manage it.

The GND proposes to manage the inherent economic transition through a large-scale economic mobilization—a coordinated deployment ("mobilization") of a nation's resources (its "economy") in response to a national crisis. Economic mobilizations organize an economy to meet needs that can be met only when *all* of a country's resources—public and private—are mobilized in line with a central, common strategy and in relentless pursuit of shared goals that supersede all other priorities. To justify an economic mobilization, a crisis must be serious enough—existential enough, really—to demand an all-out "total war effort" from both the public and private sectors.

In U.S. history, economic mobilizations have typically happened in response to a war. But that does not have to be the case. The New Deal, for example, was designed to address the Great Depression and the Dust Bowl, dual economic and environmental emergencies, similar to what we now face on a larger scale.

But every economic mobilization in American history has exploited marginalized people. Ninety-eight percent of the government-backed * home loans provided as part of the New Deal and the World War II GI Bill went to White Americans. The highway expansion razed urban communities, erasing decades of wealth for immigrants and people of color. Some New Deal employment programs, including the Civilian Conservation Corps, prohibited women and made it exceedingly challenging for Black people to participate. Because of this, for millions of Americans, the thought of an economic mobilization creates a justifiable fear—even in the face of a possible apocalypse and especially when coupled with the fallout they know will result from moving away from fossil fuels. We wanted the GND to address these fears directly.

More important, we wanted to show that an economic mobilization without a clear focus on justice and equity is, in fact, a danger—both to marginalized people and to our ability to effectively mitigate the climate crisis. The legacies of past economic mobilization have exacerbated *both* climate change and economic social injustice. For example, the racist distribution of home loans fueled redlining, residential segregation, *and* suburban sprawl, all of which exacerbate emissions. Similarly, many of the communities that were split open to build interstate highways never economically recovered and have since become frontline communities with high levels of fossil fuel pollution and low levels of air quality. That is why the GND defines the problem of climate change differently, as both a problem of government inaction and a result of government action. This is the only way to craft effective climate policy—and fuel the movement to pass it.

To transition to 100 percent renewable electricity, one study estimates, the United States will need roughly 78 million rooftop solar arrays, 485,000 wind turbines, and 49,000 solar power plants, not to mention massive energy storage—all new. Similar transformations need to happen in nearly every sector of the economy, including buildings, transportation, agriculture, and manufacturing. As in World War II, the United States has to not only equip itself but develop and produce low-carbon goods for other nations too.

As with any economic mobilization, it is not enough to simply produce the necessary technology; we must also build—and manage—the infrastructure to support it. That means millions of miles of new transmission cables to support "smart" electric grids that can integrate renewables; thousands of new charging stations for electric vehicles; and new manufacturing facilities to produce electric furnaces, water heaters, and stoves to equip our homes and businesses—and that is still the beginning. We also need new practices to replace carbon-intensive industrial and agricultural processes. Low-carbon cements, alternative refrigerants, regenerative ocean farms—we need all of it to zero out emissions, and we need it fast.

Economic mobilizations are the only times that the United States has scaled up manufacturing at anywhere near the speed that the climate crisis requires. This is due, in part, to the policy coordination that economic mobilizations make possible. There are few other times when all of the levers of government—including regulation, legislation, executive action, and procurement—work together toward a single set of goals. But it is also due, in large part, to the unprecedented levels of public investment, especially in service to initiatives that are too "risky" or big to attract private investors. That is crucial for the climate crisis. Although the United States already has most of the necessary technology, we still need some new breakthroughs—particularly when it comes to "hard to decarbonize" sectors like aviation and heavy industry.

An economic mobilization would also allow the government to use its power—both as a customer and as a regulator—to expand demand for "green" goods and to reshape markets to support them. Take solar energy as an example. To fully transition the U.S. electricity system to renewables, about 57 percent of all suitable residential rooftops need to have solar installations. But in 2020 an average-sized residential solar system will cost between $11,400 and $14,900 *after* tax credits—far more than most American homeowners can afford. It is a vicious cycle: The price of residential solar cannot decrease until more homes install panels, but more homes cannot install solar until the prices decrease. Until now, federal and state governments have done little to help outside of offering tax credits, which do very little to defray upfront costs.

Through the GND, however, government can directly increase demand by deploying legislation to require that new construction be zero carbon; funding programs that retrofit existing homes and buildings to renewable energy and high efficiency; and establishing financing mechanisms, such as green banks, that provide grants for low-income homeowners and no-interest loans for others. All of that would change the playing field for residential solar.

Economies and societies do not exist separately. Economic mobilizations open the possibility to reexamine and renegotiate our social contract—to decide what kind of country we want to carry into the future. In mobilizations—especially ones as massive as the one proposed by the GND—that means that social policy has to serve the goals and principles of the GND *and* provide the support necessary to maintain an economy at nearly full employment. That requires changes not only to labor policy but to every part of our social safety net, from healthcare and childcare policy to workforce and housing policy.

Mobilizations create tight labor markets that, in turn, increase wages and reduce inequality. But they are not inherently just—especially if not everyone can access the jobs they create. That is why the GND includes commitments to a federal jobs guarantee, universal childcare and healthcare, and significant investments in education and training. How can we attract enough workers for a national climate mobilization if the average cost for a year of childcare ranges from $5,500 to $25,000? How can families move to better-paying jobs in the mobilization if they are dependent on employer-sponsored healthcare or, worse, cannot exceed their Medicaid asset limit lest they be disqualified? How can people reenter the labor force if they do not know where to go for training or job placement? How?

How we design and implement this next mobilization matters—for the success of the GND and for the future of our country. The GND must be the most just economic mobilization in U.S. history if we are going to succeed in averting the climate crisis and reversing the unraveling of our economy. The price of national progress cannot be exploitation and systematic oppression—not again, not unless we want to fuel the crisis that we are trying to avert. We move together, or we risk not moving at all. *E pluribus unum*, indeed.

3. REFRAME

THE PROBLEM OF THE CLIMATE CRISIS IS A FAILURE OF LANGUAGE

THE RIGHT WORD WOULD BE AS LONG AS THE LIVES OF MY GRANDCHILDREN

AND THEIR GRANDCHILDREN

MADELEINE JUBILEE SAITO

How to Talk About Climate Change

KATHARINE HAYHOE

Gazing at the Andromeda galaxy through binoculars with my science-teacher dad is one of my earliest memories. And the more I learned about science, the better it got. Who wouldn't want to know why the sky is blue, that polar bears have black skin and translucent fur, and how tiny amounts of heat-trapping gases in our atmosphere serve as the thermostat for the planet?

Despite my fascination with the universe and this planet, I still thought of human-caused climate change as a distant, far-off issue, something that only really mattered to Al Gore or those polar bears. It wasn't until I was looking for an extra credit to round out my astronomy and physics degree at the University of Toronto and I ended up in a class on climate science that my perspective abruptly changed. In that course I learned it's not the planet itself—which can feel so vast it is easy to detach from—that is most at risk from a changing climate; it is all of us who call it our home.

I switched fields and headed to grad school at the University of Illinois at Urbana-Champaign to study climate science. For the past twenty years, I've been working with cities, states, and federal agencies to figure out how to prepare for the impacts of a changing climate. And I don't just study climate change—I also talk about it. A lot. In classes I teach in person at Texas Tech University and online around the world. On Twitter and Instagram and Reddit AMAs. To farmers and oil-and-gas executives and congressional staffers. On local news and the *Today* show. I'll talk just about anywhere and to just about anyone.

But the more I talk about it, the more pushback I get. I'm accused of lying and peddling "UN-derived satanic deception" and a multitude of other sins when I'm up front about what the data tells us: Climate is changing, and humans are responsible. "Global warming is a freaky genocidal doomsday cult," one man tweeted at me. "You lie for money," another posted on Facebook, "and change the data." I've been called a loony, a fraud, and a clown.

Yet as vocal as such people are and as much as they dominate the discussion, they're only a small proportion of the population. * Seventy-three percent of people in the United States agree the planet is warming, according to a poll conducted by the Yale Program on Climate Change Communication. And 62 percent of Americans recognize that the main reason for this warming is human activity: specifically, burning fossil fuels (that's about three-quarters of the problem) and deforestation and agriculture (that's most of the other quarter).

Our biggest problem isn't skeptics who perpetuate the idea that science is somehow optional or a matter of opinion—it's that when it comes to supporting climate action, the urgency just isn't there for * many of us. Seventy-three percent of us also believe climate change will affect future generations, but only 42 percent think it will affect us in our lifetimes. As for the solutions, we're told—incorrectly—that they're expensive and ineffective, and so we wonder why we should bear the brunt of the financial impact.

The reality, though, is that climate change is affecting us today. And it's doing this by taking many of the risks we already face naturally—floods and storms, heat and drought—and supersizing or exacerbating them. This isn't about saving the planet. The planet itself will survive. The question is what will happen to the rest of us who live here.

That's exactly why talking about climate change is so important. If we don't talk about why it matters, why would we care about the problem itself? And if we don't talk about what we can do to fix it, why would we take action or expect our community, our state, and our country to do so either? As challenging, stressful, and painful as it might be, addressing climate change begins by actually talking about it. And over the years, I found a way to do so that's actually constructive. It begins with why climate change matters to us.

Why It's So Urgent

* The decade that just closed was the warmest on record, both across the globe and in the United States. Around the country, climate change is already leading to more frequent heavy-rain events and

record-breaking heat waves, increasing the area burned by wildfires, and supersizing our storms and hurricanes. As the planet warms, it's loading the weather dice against us: making some events stronger, others more intense, still others more frequent, and nearly all of them more destructive. It is estimated that almost 40 percent of the rain that fell during Hurricane Harvey, for example, was the direct result of human-induced warming. The area burned by wildfires across the western United States has more than doubled since the 1980s as a result of a warmer world. "Bomb cyclones" in the Midwest and atmospheric rivers along the West Coast are intensifying as the climate changes, increasing the risk of devastating floods; and Miami Beach is trying to stave off the impacts of sea level rise by raising its streets, even though experts say what's planned is still not enough.

And we're the fortunate ones: The impact of these supercharged weather extremes on poor countries far exceeds what we see in North America. Since the 1960s, it's estimated, climate change has widened the economic gap between the richest and poorest countries in the world by as much as 25 percent.

My specialty is high-resolution climate projections: translating our big, global climate models into information that shows how an individual place (such as a city, state, or broader region) will be affected by a changing climate. As a scientist, my job has always involved a lot of late nights coding on a computer and writing detailed descriptions of my research for scientific publications. But as interest in climate change grows, more of my time is being spent answering people's questions: When will my family's farm run out of water? What risks does climate change pose to our city? How can we transition our energy systems off fossil fuels without harming the economy here or development abroad?

I've come to realize that responding to these questions is just as important as the science side of my job. Scientists can't expect to change the world from behind a computer screen, no matter how many reports we publish each year or how long they are. (The most recent U.S. National Climate Assessment, which I helped write, clocked in at more than two thousand pages.) It's increasingly urgent that we understand the tangible impacts climate change is having on our lives today and how it affects the things that matter to us.

A few years ago, I was invited to speak at the Successful Canadian Women's Dinner. It's a fundraising benefit for Adsum for Women and Children, a nonprofit that supports women and families experiencing homelessness in Halifax, Nova Scotia.

I'd spent that day traveling around the city with Sheri Lecker, executive director of Adsum. She shared how the previous summer's record-breaking heat had driven more people to seek shelter. Severe rainfall also made it harder to arrange transportation to appointments and jobs because bus routes were shut down or delayed, and they also had to deal with what happened when people missed medical appointments and counseling programs. The implications a changing climate has on Adsum's work is clear, as is the organization's dedication to women and kids, the very people who are disproportionately affected by climate change and the increasing risk of weather-related disasters around the world.

I put all this into my talk. When I finished, one of the sponsors was the first to grab my hand. "I have to admit, I wondered what they were thinking when they invited you," he said. "But that was the best talk we've ever had!" He explained it had helped him connect the dots between climate change and what mattered to him—people. And through doing so, he'd recognized the most important truth of all: It isn't a matter of moving climate change further up our priority list. The reason we care about it is because it already affects everything that's at the top of our priority list: our health, our families, our jobs and the economy, the well-being of our communities and those less fortunate than us who live in them. To care about a changing climate we don't have to be a tree hugger or an environmentalist (though it certainly helps); as long as we are a human alive today, then who we already are, and what we already care about, gives us all the reasons we need.

Why We Need to Find Common Ground

I learned that important truth from the first conversation I had with someone who disagreed with me on climate change—my husband.

I was vaguely aware, during my first few years of studying climate change in the States, that there were people who didn't think climate change was real. But in the United States today, our opinions on climate change are based on our politics, not our knowledge. So my husband, who grew up on a horse farm in conservative Virginia, had never met anyone who thought it was happening, and certainly no one who thought humans were responsible.

It sounds daunting, but we had two big advantages: a lot of shared interests and a lot of motivation to work this out. Over the next year, we had dozens of conversations, some sitting side by side at the computer and looking at global temperature data from NASA, others talking about what happens to people's jobs when we stop using coal. We're now on the same page when it comes to this issue, but this experience taught me how critical it is that we begin these discussions with mutual respect and a focus on what genuinely connects us.

Today when I encounter someone who's doubtful about the reality or the relevance of climate change, I don't start by talking science; instead, I try to identify something we have in common. If they're a skier, it's important to know that the snowpack is shrinking as our winters warm; maybe they'd like to hear more about the work of an organization like Protect Our Winters, which advocates for climate action. If they're a birder, they might have noticed how climate change is altering the migration patterns of birds; the National Audubon Society has mapped future distributions for many native species, showing just how radically different they'll be from today. If they're a parent, like me, we both know how worrying it is that our children are living in a world that is far less stable than the one we grew up in.

A few years ago, I was invited to speak at the Rotary Club in West Texas, where I live. As I walked in, I noticed a giant banner stating the Four-Way Test, the Rotarian's ethical guidepost: "Is it the truth? Is it fair to all concerned? Will it build goodwill and better friendships? Will it be beneficial to all concerned?" I'm not a Rotarian, but these values hit me right in the eye.

Is what we know about climate change the truth? Yes, it absolutely is. We've known since the 1850s that digging up and then burning coal—and, later, oil and gas—produces heat-trapping gases that are wrapping an extra blanket around the planet. Since then, thousands of studies and millions of data points have confirmed it's true. Together with colleagues from Norway and Australia, I've even taken the few dozen studies that suggest this isn't the case and recalculated their work from scratch. In each, we found an error that, when corrected, brought the results right back into line with the thousands of studies that agree the climate is changing, humans are responsible, and the impacts are serious.

Is climate change fair? Absolutely not. The poorest and most vulnerable among us, those who have done the least to contribute to the problem, are most affected. These include the women and children Adsum supports in Halifax; farmers struggling to raise their crops in East Africa; Bangladeshis losing their land to sea level rise and erosion; and Arctic peoples whose traditions are threatened and whose homes are being displaced by rising seas and thawing permafrost. The carbon footprint of these groups is minuscule. They've contributed so little to the problem, yet they bear the brunt of the impacts. The eighty-five lowest-emitting countries in the world, the Climate Vulnerable Forum estimates, who have contributed virtually nothing to the problem, will bear 40 percent of the economic losses and 80 percent of the resulting deaths from human-induced climate change. That is absolutely not fair.

And would it build goodwill and be beneficial to address climate change? Yes, it would. The more the climate changes, the more serious and ultimately dangerous its impacts become. In Texas, climate change is amplifying our natural cycle of wet and dry, making our droughts stronger and longer at the same time it supercharges hurricanes and extreme rain. My research shows that the sooner we cut our carbon emissions, the greater and more costly the impacts we'll avoid. And transitioning to clean energy brings new tech and opportunities as well, including around 35,000 jobs in Texas already. As we work together, we can build goodwill.

I ignored the Rotary lunch buffet and instead whipped out my laptop. As fast as I could, I rearranged my presentation on how climate change affects West Texas to track the Four-Way Test. I was glad I did, because when I stood up to speak, I could see many more skeptical faces than I saw at the Adsum banquet—people who questioned not just climate change's relevance to their lives but its reality in the first place. But as I spoke, I could see those faces changing and heads nodding. And I will never forget the local banker who had the final word: "I wasn't too sure about this whole global warming thing, but it passed the Four-Way Test!"

I had persuaded him by beginning with his values, showing my respect for them, and then connecting the dots between what he already cared about and a changing climate. And it worked—because to care about climate change, all we really have to be is a human living on planet Earth.

She Told Me the Earth Loves Us

ANNE HAVEN MCDONNELL

She said it softly, without a need
for conviction or romance.
After everything? I asked, ashamed.

That's not the kind of love she meant.
She walked through a field of gray
beetle-bored pine, snags branching

like polished bone. I forget sometimes
how trees look at me with the generosity
of water. I forget all the other

breath I'm breathing in.
Today I learned that trees can't sleep
with our lights on. That they knit

a forest in their language, their feelings.
This is not a metaphor.
Like seeing a face across a crowd,

we are learning all the old things,
newly shined and numbered.
I'm always looking

for a place to lie down
and cry. Green, mossed, shaded.
Or rock-quiet, empty. Somewhere

to hush and start over.
I put on my antlers in the sun.
I walk through the dark gates of the trees.

Grief waters my footsteps, leaving
a trail that glistens.

Truth Be Told

Emily Atkin

The climate crisis is, in part, a failure to respond to information. We've known for decades that there would be catastrophic consequences if we continued to burn fossil fuels. But we haven't truly accepted that fact or responded to it. Why?

One reason is the information itself. The fossil fuel industry has lobbed misinformation and incited public confusion, and the media hasn't helped. By and large, mainstream journalism has failed to meet the challenges of the climate crisis. Citizens simply haven't had the quality nor the quantity of information they need to respond adequately to the problem. Thus, there's been an ongoing debate on how journalists should—and should not—cover climate change.

My winding path as a journalist is a reflection of that debate. Through it, I've developed a strong sense of what needs to change to get this story right.

My professors taught that journalism was essential to democracy. Voters could make informed decisions to solve society's most complex problems only if they were adequately informed. If voters weren't making smart decisions, that simply meant journalists weren't doing their jobs.

I loved this theory so much that I decided to shape a career around it. An investigative reporter named Wayne Barrett took me under his wing. If any journalist knew the importance of reporting, it was Wayne. A more-than-thirty-year veteran of *The Village Voice*, Wayne had been one of the first journalists to uncover the self-dealings of powerful men like Rudy Giuliani, Al Sharpton, and Donald Trump. His drive to afflict the comfortable and comfort the afflicted was so intense it was almost frightening. Once, while doing some reporting for Wayne, I forgot to ask a key question of a campaign operative, and Wayne reprimanded: "Readers deserved this information. Now they won't have it."

Even with Wayne's help, though, I couldn't land a job in the ultra-competitive world of political journalism. After two years of failure, I decided to break into the industry another way and applied for a position in Washington, DC, reporting on climate change.

Wayne thought climate reporting was a worthwhile pursuit. After all, climate change was one of society's biggest and most complex problems—and people deserved to know about it and decide on how to fix it. Climate reporting also seemed like a unique opportunity to make a difference, because climate change was a solvable problem.

I got the job in November of 2013. It was at ThinkProgress, a now-defunct news website run by the policy think tank Center for American Progress. My very first assignment was about a prospective Texas Senate candidate who said global warming was God's punishment for women who got abortions. Obviously, this was wrong, and this guy was a dangerous idiot. But I kept those thoughts to myself. When I started reporting on climate, I was worried that because I was associated with a liberal think tank, people wouldn't trust me; therefore, I vowed to stick to the facts. After all, as Wayne had once written: "The joy of our profession is discovery, not dissertation."

"The most recent report issued by the Intergovernmental Panel on Climate Change confirmed that 'it is extremely likely that more than half of the observed increase in global average surface temperature from 1951 to 2010 was caused by the anthropogenic increase in greenhouse gas concentrations and other anthropogenic forcings together,'" I wrote, adding: "The report did not mention abortion."

Over the next two years, I fact-checked dozens of instances of climate misinformation without passing judgment on those who lied. I explained myriad terrifying scientific studies without explicitly remarking on how terrifying they were. I reported from the front lines of environmental injustices all over the country without saying the victims deserved better. My goal was that, if you took my writing and slapped Associated Press branding on it, you wouldn't be able to tell it was from a "left-leaning" publication.

But it turns out staying neutral isn't easy when you're not actually neutral. Because the reality was I didn't want climate change to get worse, and I didn't want people to suffer. Every time I didn't say that, I felt like I was the one lying.

I remember the first time I felt legitimately alarmed. It was in 2014, after writing about a World Meteorological Organization report showing carbon accumulating in the atmosphere far more rapidly than expected. Once carbon concentrations reached a certain point, the report said, the subsequent warming would trigger feedback loops of more carbon releases and more warming, causing unpredictable levels of suffering for the world's poorest and most vulnerable populations. "We must reverse this trend," the WMO's secretary-general said. "We are running out of time."

I felt even more alarmed as I reported on the willful ignorance of those tasked with solving this looming crisis. Instead of preparing for climate change, state government agencies were removing scientific information about it from their websites. Instead of trying to stop climate change, politicians were making up ridiculous excuses about why it didn't exist—like that global warming couldn't be bad, because Mars was warming too.

Worst of all, no one seemed to be paying attention. At least not to my reporting. It was like watching from a crowded beach as a child drowns, everyone ignoring their screams for help.

In August of 2015, I cracked. Frightened about the climate crisis, pessimistic about the future, and self-conscious about my own ineffectiveness, I asked my boss if I could switch to politics. Perhaps there I could make Wayne proud.

For the next eleven months, I traveled the country covering the presidential election, trying to reignite the spark inside me that climate reporting had dimmed.

At first it was exactly what I needed. Political reporting was more fast-paced and competitive, and the subject matter was more wide-ranging. In my first trip to Iowa alone, I covered Ted Cruz's promises to be antiestablishment; women's-health activists throwing condoms at Carly Fiorina; and a Republican college student confronting Marco Rubio about climate change.

Best of all, people seemed to care. Traffic-wise, my stories did well. I even got some nice emails.

The high, however, was short-lived—because I quickly realized clicks aren't a real measure of societal impact. This was evidenced by the fact that Donald Trump was almost surely about to clinch the Republican nomination for president. Wayne had spent years exposing Trump as a corrupt, self-dealing real estate industrialist in New York. Other journalists and I had spent months exposing him as a climate denier and a threat to a livable planet. So why did people still want this person to be president? I couldn't understand it.

I convinced myself it was because much of the political reporting style was too fast-paced, too surface level, too targeted to an already-liberal audience. So in July of 2016, just as Trump became the Republican nominee, I took a political reporting job at Sinclair Broadcast Group, which was known for being conservative leaning and had stations in swing districts all over the country.

This position felt like a true opportunity to move the needle on public understanding of many issues—climate change in particular. For months I worked on a segment about how sea level rise was threatening one of the nation's most important air force bases, near Norfolk, Virginia. I got several high-level military officials on camera explaining the threat and interviewed a conservative think-tanker who admitted that sea level rise was a worthwhile threat for the military to tackle.

I was so proud of that segment. A long version of it aired online, and a shorter one on some of Sinclair's TV stations. I waited for viewers' reaction. It never came.

When Trump was elected president, I was standing in Times Square doing a Facebook Live video about how tourists were reacting to the news. It felt just as pointless as the climate segment and everything else I had ever done in journalism. Fear washed over me about the future—not just of the climate but of democracy. For the first time in my adult life, I questioned the path I had chosen and the theory of change I had shaped my life around.

A few months after the election, I decided to quit Sinclair. I emailed Wayne to tell him and to ask him what I should do next. But he never responded. Two weeks later—the day before Trump's inauguration—Wayne passed away from interstitial lung disease.

Wayne's funeral was so packed you could barely move. Reporters he mentored from all over the country were there, as were politicians from all over the city and state. Andrew Cuomo and Chuck Schumer each spoke at the funeral about how much they feared Wayne—and how much they respected him. They said his anger and relentlessness kept them honest. They said society needed more journalists like him.

That day, I realized something about Wayne I hadn't before. He may have been a reporter at heart, but he never did it with a straight face. If he uncovered facts that showed a politician was a liar, he called them a goddamn liar. If he uncovered facts that showed a business executive was self-dealing, he called them a grifter.

Wayne did more than just uncover the facts. He told the truth.

I knew the truth about climate change. And the truth was we were running out of time. The truth was our new president and the fossil fuel industry that had helped elect him seemed content to allow millions of people, animals, and ecosystems to suffer and die. Those victims included not just the most vulnerable people in society but me and the ones I loved most. Wayne wouldn't have been "objective" about this. He would have taken the bastards to task.

I decided that this was how I was going to approach reporting from then on. With the climate crisis accelerating and a climate denier in the White House, there wasn't time to waste. I applied for a job covering climate change at *The New Republic*, a journal of reported opinion, and got it.

It was incredibly difficult to learn how to speak my mind on the page. I remember feeling so uncomfortable when, in May 2017, I wrote an article calling Scott Pruitt, the now-disgraced former head of the EPA, a "hypocritical liar." But my editor assured me that this was what the reporting of the facts about Pruitt showed. Pruitt had spent his whole career suing the Obama administration's EPA to prevent it from taking action to protect the environment and now he was criticizing Obama for not having done enough to protect the environment. This was hypocritical. Pruitt claimed he was going to do more to protect the environment than Obama had, but Pruitt was already dismantling environmental regulations and denying climate change. This made him a liar.

With time, I started to feel confident about being a moral arbiter—that is, a journalist who calls it out when someone is doing something morally reprehensible, like prioritizing money over protecting the vulnerable. And I started to feel more fulfilled.

I was also starting to feel angry—angry at all the politicians, corporations, and other powerful interests preventing progress who were content to doom the future in favor of short-term gain. I wanted to report exclusively on what made me most angry about climate change: the "bosses, the wire pullers, the campaign givers and takers," as Wayne would say. Those with the money and power to demand and achieve systemic change who choose to instead sit on the sidelines and watch from high ground as vulnerable people suffer.

I also suspected I wasn't the only person who felt this way and who wanted these things. I wanted to connect with those people more directly. I remember how, at *The Village Voice*, Wayne wasn't writing to everyone in the city. He was writing to the paper's niche, progressive audience. It wasn't his voice alone that moved the needle. It was their voices—his and his audience's—together.

So in September 2019, I left *The New Republic* to start an independent publication, HEATED.

I chose a newsletter format because it felt more intimate—a way to speak to readers in a conversational voice. The topic of climate change was both serious and confusing enough on its own; it didn't need to be bogged down by fancy Latin phrases or four-syllable adjectives.

Almost immediately, I started to see the type of impact I had dreamed my words could make. In HEATED's first month of operation, I named the companies that were advertising on *The Michael Knowles Show*, which is hosted by a climate denier who called Greta Thunberg "mentally ill." Following publication of that article and the ensuing backlash, Vistaprint announced it would not be advertising "on any upcoming episodes of the Michael Knowles podcast, now or in the future." A few weeks later, HEATED published an investigation into how Twitter's upcoming ad policy would benefit fossil fuel companies while harming climate advocacy groups. The piece sparked a national conversation about Twitter's ad policy and climate change—in part thanks to being tweeted by Senator Elizabeth Warren—and ultimately culminated in Twitter changing its ad policy.

The most important impact HEATED is having, however, seems to be with readers. In November I asked them to tell me why they enjoy or don't enjoy the publication. I got 111 responses and compiled them in a spreadsheet. Three-quarters of readers said they enjoyed the shift in tone from dispassionate to passionate. Six people explicitly said it made them feel "less alone." "Believing in climate change can feel isolating sometimes, but we're not alone in this fight, and your newsletter helps me remember that," one reader wrote.

For years, I assumed that educating people who are already interested in climate change would just be preaching to the choir. But my reporting career has taught me that's not true. Most people who are interested in climate change just don't yet have the tools to talk about it confidently. The choir is there. They want to sing. But they don't know the words.

My journalism professors taught me our job was to provide those words—to give citizens the information they need to solve society's most complex problems. Wayne taught me our responsibility was to do that fearlessly, with righteous anger on behalf of society's most vulnerable. Wayne's mentor, the legendary *Village Voice* journalist Jack Newfield, said it best: "Compassion without anger can become merely sentiment or pity. Knowledge without anger can stagnate into mere cynicism and apathy. Anger improves lucidity, persistence, audacity, and memory."

To stop the forces that have been preventing climate action for more than three decades, we need journalism that embodies all those things. And the stakes of getting that right could not be higher. We have only so many opportunities to get it right before everything goes horribly wrong.

Today, mainstream climate journalism is improving—but not nearly fast enough to keep up with the severity of the crisis. So it is not just up to our journalistic institutions to demand seriousness, righteous anger, and relentlessness from themselves. Every climate-conscious news consumer should demand it too. Our journalism outlets purport to serve the public by giving us the information we need to survive. If we don't collectively hold media's feet to the fire, our own toes will be the ones that burn.

The function of art is to do more than tell it like it is—it's to imagine what is *possible*.

—BELL HOOKS

Harnessing Cultural Power

FAVIANNA RODRIGUEZ

To the leaders, organizers, and funders of the climate movement:

Culture is power. The music we listen to, the social media we consume, the food we eat, the movies and television shows we watch—these all inform our values, behaviors, and worldviews. Culture is in a constant battle for our imagination. It is our most powerful tool to inspire the social change these times demand. Current global events, from pandemics to massive wildfires, demonstrate how interconnected and interdependent we are as human beings, with one another and with Earth. As the old narrative of capitalism reveals its devastating failures, we urgently need more compelling and relatable stories that show us what a just, sustainable, and healthy world can look like. The old myths will die when we can replace them with new ones. We need our storytellers—a mighty force—to help us shift our mythology and imagine a future where together we thrive *with* nature. That is a power we must harness, if we are to find our way out of the climate crisis. We must build a cultural strategy for the climate movement.

Perspectives and points of view shape societal values, practices, and behaviors. They influence our understanding of everything from gender to race to our relationship with the natural world. Our current relationship to the Earth is based on a worldview of domination that supports an extractive economy. This is a myth of man's making, and it's one that has influenced our cultural imaginations since Westerners conquered the land, ravaged Indigenous communities, and built a society around fossil fuel extraction, industrial animal agriculture, cheap labor, and what Greta Thunberg calls "fairy tales of eternal economic growth."

Though I didn't quite know it at the time, I started to learn these lessons as a kid, growing up in a concrete jungle. Highly creative, I loved to draw and paint and create new worlds on my terms, but my family was busy surviving. My parents, immigrants who migrated to

California from Peru in 1968, raised me in East Oakland—a neighborhood of dead cement and abandoned industrial buildings—at a time when the crack epidemic was ravaging my community. Police brutality and gang violence surrounded me. But it was also a place with a strong legacy of the unapologetic culture of the Black Panther Party, an example of how the status quo can shift and community support networks emerge when people create culture *on their terms* to confront dominant power structures.

Today I can see the invisible force that was ravaging my childhood community—the pollution in my hood that was causing asthma and making my neighbors sick. I lived next to the I-880 highway, which carries the highest volume of truck traffic in the San Francisco Bay Area as it slices through communities of color, including the neighborhood I still call home. In contrast, the nearby I-580, which cuts through more affluent and White communities, did not allow trucks. Those communities were protected, while we ingested toxic diesel fumes that cut our life spans.

I didn't have the language back then to connect the dots, but I do today. My community, like most communities of color and Indigenous communities, was intentionally disenfranchised, poisoned, and turned into a dumping ground. Injustices intersected: violence, poverty, drugs, government neglect, and environmental racism.

When it comes to climate change, most of the stories and cultural content that exist to inform and organize people are overwhelmingly pain oriented, outdated, and *hella* White. They don't reveal the racial and economic drivers behind the climate crisis; they don't reflect the reasons why my community was exposed to pollution while White communities just a few miles away had better air quality. In fact, most often they either leave my people out of the story entirely or center White men as activist heroes and rely on doom-and-gloom narratives.

We are not seeing diverse stories about climate because the folks who control the cultural engine in the United States—TV, film, visual art, music, gaming, performance, publishing—are overwhelmingly White men. And they exist in a status quo that has entrenched their power and dramatically narrowed our collective gaze, including how we view our relationship to the natural world and our practices for caring for one another. Just as ecosystems need biodiversity to thrive, society needs cultural diversity to grow new possibilities. Monoculture deadens our collective potential.

Not only are we not seeing diverse stories, but we're not seeing enough stories, period. A recent look at episodic TV shows found that in 2019 only three dealt with climate change (excluding docuseries that explicitly focus on climate). *Vice* noted, "A crisis that's reshaping every aspect of human experience is being effectively ignored by TV."

The power of culture lies in the power of story. Stories change and activate people, and people have the power to change norms, cultural practices, and systems. Stories are like individual stars. For thousands of years, humans used the stars to tell stories, to help make sense of their lives, to orient them on the planet. Stories work in the same way. When many stars coalesce around similar themes, they form a narrative constellation that can disrupt business as usual. They reveal patterns and help illuminate that which was once obscured. The powerful shine in one story can inspire other stories. We need more transformational stories so that we can connect the dots and shift narratives.

The climate movement has largely left storytellers and culture out of its strategy toolbox. Now is the time to change that. Here's how we can harness the power of culture for climate action.

Pass the mic to artists and culture-makers of color.

Most of the prominent cultural icons speaking out about climate change are White. This is not because people of color don't get it or don't care—in fact, polling shows people of color are more alarmed or concerned than their White counterparts. That includes 69 percent of Latinx Americans, the most of any racial group. Gatekeepers must welcome in the wisdom of the communities most impacted by the extractive economy—frontline communities of color and Indigenous communities. Centering these storytellers—in television, in print media, and in pop culture—can mobilize important constituencies, particularly those who will be directly affected by climate chaos and displacement but who do not relate to the mostly White actors, politicians, environmentalists, and scientists who are the planet's current high-profile spokespeople. Imagine hip-hop songs that say "F*ck the polluters" or Indigenous storytellers being nominated on every Oscars slate. Culture anchored in the voices of those marginalized by our current command-extract-control society is critical for the urgent transformation that's required.

Build diverse cultural infrastructure.

Right now, there are virtually no comprehensive support systems to help creators, musicians, screenwriters, poets, journalists, and artists—those who shape the culture we consume every day—to engage effectively with climate issues. This is too critical an issue to leave everyone grappling with it on their own, expecting individuals to educate themselves and find their own way to engage. To fill that gap, I founded the Center for Cultural Power to support culture makers as they engage more deeply with climate topics and other social justice issues and to harness their immense power to shape our core narratives. Imagine trainings and fellowships for artists and writers to learn about the climate crisis; funding to create new content that is intersectional and speaks to multiracial audiences; more support for artists of color, who are already facing barriers in the cultural industries because of the domination of White men in power. By fostering a more equitable cultural sector, we can cultivate the cultural content that propels robust climate action.

Include artists and makers in your work.

We need a new wave of artists who are succeeding in making visible the stories of those most affected. We must include creatives more thoughtfully in our change work, not just through one-off transactions—like having a musician at your gala—but through long-term relationships with artists that build our cultural capacity. This can look like hosting an artist-in-residence, collaborating with artists to facilitate storytelling programs for your constituents, inviting an artist to join your board, teaming up with artists to design protests, or commissioning cultural activations. For example, during the 2019 Global Climate Strikes, a coalition of artists and climate groups organized a massive live-painting intervention consisting of eleven twenty-five-foot murals in the center of San Francisco's financial district, an area lined by San Francisco's biggest banks. The murals featured colorful solutions to climate chaos, with messages like "Green New Deal" and "Return to the Old Ways," and hundreds of community members participated in painting them. The urgency of the climate crisis demands that we develop strategies that embrace the enthusiasm of artists and nurture an abundance of cultural and narrative strategies to reimagine life on Earth.

Make human stories to move human beings.

Human stories are more powerful for inciting action than counting carbon or detailing melting glaciers. Fossil fuel extraction is not just a lamentable occurrence. Oil does not extract itself; gas doesn't burn on its own. To talk about ecological devastation and climate chaos, we must talk about the fact that we live in a culture that condones the exploitation of other human beings and the lands on which they live and severely impacts communities, whether in East Los Angeles or on tribal lands in Standing Rock. For example, I've been deeply inspired to see how girls in the climate movement have rapidly popularized a new story about the responsibility that adults have to young people and have spoken with unapologetic fearlessness in confronting the establishment. We must challenge the idea that some life matters more than other life. This means creating more stories that center the experiences of human beings, showcase solutions, activate our human empathy, and help usher in another worldview.

Create culture to challenge consumption.

For centuries, we have been told a story that sources of energy are endless, that extraction of fossil fuels can continue without bounds. While we know we need massive expansion of electrified public transportation to transition from fossil fuels, how many awesome movies do you see that center the individual car as a status symbol? While we know the planet cannot sustain the level of meat consumption that is devastating our world's forests and exploiting the animal kingdom, how much of our food culture has normalized flesh on our tables?

Where is the pop culture that makes riding public transportation and eating a plant-based diet fun, cool, and accessible to diverse audiences? Imagine the power of being exposed to an abundance of stories, songs, and images that challenge our fundamental consumption culture and expand our perspectives by helping us *feel* the consequences of our choices. What if we made it uncool to use fossil fuels in the same way smoking became uncool?

Let culture connect us to nature.

A global culture of domination and extraction has severed our human relationship to nature, but a culture of stewardship and ecological harmony can reconnect us to our ancestral stories in which we were more deeply connected to nature and her rhythms. Culture allows us to confront, acknowledge, and mourn what was lost, while offering a way to move forward. The forced displacement and enslavement of African people, the genocide of Indigenous people, and the colonization project more generally have had a devastating impact on our relationship to nature, and that is a wound that must be tended. Consider the power of Indigenous storytelling in the fight against fossil fuel corporations. During protests at Standing Rock against the Dakota Access Pipeline, the story of the evil black snake and the heroic water protectors was so powerful that it united Native tribes and activated people of all races from all walks of life. Many Indigenous communities hold a worldview that the land is not ours to exploit—rather, that we are stewards of nature. A climate-movement strategy should make visible the work of Native and Black communities. Stories can decolonize our imaginations.

Stand in the power of *yes*!

In the social justice movement, I've observed that our work is often centered on what we are against. We are clear about what we don't want—the *no*. And that's understandable when our communities are constantly being attacked. Our movements become our first line of defense. But we cannot envision a future when we're stuck in fight or flight. We must also create a culture that is about our *yes*, and this is where we can rely on artists. For example, artist Molly Crabapple, writer Naomi Klein, and filmmaker Avi Lewis collaborated with Alexandria Ocasio-Cortez to create a powerful illustrated video about what the future could look like if we have a Green New Deal in the United States. This piece of art, grounded in AOC's personal story, went viral and is one of the few pop culture pieces to date about the future that's possible with a Green New Deal. Imagine an outpouring of cultural content that shows us a future where political, economic, and cultural power are justly distributed and humans are in a regenerative relationship with nature. There must be room for creation that is captivating and irresistible. We can use our radical imagination to visualize and manifest another world, and we can make that world feel real through cultural products such as TV shows, films, comic books, images, or songs.

The stories we tell will determine whether our society declines and self-destructs or whether we can heal and thrive.

Remember that every story is based on a particular perspective on the world. We must always consider how an author's point of view affects the myths that form our reality. The Nigerian novelist and poet Chinua Achebe got it right when he said, "Until the lions have their own historians, the history of the hunt will always glorify the hunter." Let's ask ourselves: Who is telling the story or creating the image? What values do they espouse? Whom do I see? Whom do I *not* see? What worldview is being communicated? How is nature treated? Who stands to gain from this way of seeing the world? Whom do they blame for the problem? Who would benefit from the solutions they offer?

As I stand in my power as an artist and climate justice leader, I now understand that it's time to write a new story. I cannot heal my community or myself without healing the planet, and we cannot save the planet without healing injustice. So the question is: Will you stand with me in harnessing culture for the betterment of Earth, to save life as we know it?

With love and an unbounded imagination,
FAVIANNA

Leaderful means there is enough room for all of us. Seeing everyone roll in together is much more powerful than having one or two people speak for everyone. Being inter-generationally leaderful also generates the best ideas and solutions. . . . We need to do this together, and we can do it lovingly.

—Elizabeth Yeampierre

Becoming a Climate Citizen

Kate Knuth

As I try to make sense of how to navigate this time, I've come to see "climate citizenship" as a path forward. This concept asserts that any chance of society gracefully navigating the climate crisis will take a renaissance of citizenship and civic life. Likewise, the scale and scope of the climate crisis mean that it will reshape citizenship in fundamental ways.

Though I didn't call it climate citizenship at the time, this idea has its roots in the earliest days of my career. In August 2005, I was completing a master's degree in conservation at the University of Oxford. I spent my days toggling between editing my thesis and reading news coverage of Hurricane Katrina and its aftermath. The storm in motion was horrifying, and the days that followed were gut-wrenching. For the first time, I understood that climate change would not just unleash dangerous weather; it was going to rip its way through all of society's imperfections, laying bare our unjust systems in ways that would leave people dead. In fact, it was already killing people. And it was starting to show the potential to wipe out entire cultures.

As I finished up my schooling and Katrina revealed the power and injustice of climate change, I asked myself: *What's the best thing I can do in response?*

My answer was to move back to my hometown of New Brighton, Minnesota—a suburb of Minneapolis and Saint Paul—and recommit to my roots. Six weeks later, I launched my campaign for the Minnesota House of Representatives.

The next year gave me a deep education in democracy. I realized early on that running for office involves a lot of listening. After connecting with the most active Democrats in my area and securing a contested party endorsement, I turned my attention to a critical central task: knocking on doors and listening. A single father told me of struggling to pay for both his own insulin and his son's epilepsy medication on a butcher's salary. An elderly man described living through the Great Depression and his gratitude for the Social Security program created during that time. One summer night, a group of neighbors shared freezies with me, and while their kids played around us, we talked about a clean energy future. My constituents taught me that democracy depends on people's willingness to share their hopes and fears with one another in order to weave a better future, together.

With lots of help, I won that election. In January 2007, I raised my right hand and swore an oath to uphold the constitutions of Minnesota and the United States. I also swore to myself to find ways to uphold the planet's living systems.

The next two years were heady ones for working on climate. In Minnesota we passed a nation-leading renewable-energy standard with bipartisan support, and I got to work on a Minnesota cap-and-trade bill—a market-based approach for controlling emissions. Minnesota, led by a Republican governor, joined the Midwestern Greenhouse Gas Reduction Accord. Maybe, I found myself imagining, *just maybe* we'd have a bunch of states in the middle of the country with a price on carbon. Then, in 2009, a bill to curb greenhouse gas emissions actually made it through one house of the U.S. Congress. I genuinely believed that our politics, and our democracy, could handle climate change. It might be a little bumpy, but we would handle it.

I was wrong—though it took me a few years to realize my miscalculation. There is, in fact, no certainty that our democracy can handle climate change. However, I'm still convinced our only hope for successfully grappling with this crisis will come from deep investment in the day-to-day work of democracy and, more specifically, citizenship.

We are living in a time of grief for lives and species already lost to a changing climate; a time of anxiety about the uncertainty of our own futures and those of our children; and a time of opportunity to capture the benefits of a huge, fast energy transition.

At the root of it all, we are living in a time of transformation. Given the emissions already in the atmosphere, and those predicted to come, especially without major change, there is no question the world will be reshaped. This is more than a fiddling-around-the-edges sort of change. It's a fundamentally-different-system-emerging change.

The early stages of transformation—where we find ourselves

today—feel unsteady. They feel like the in-between mind of a teenager, neither child nor adult. For me right now, that unsteadiness feels like the sting of tears welling up as I sip lukewarm decaf coffee and read news of yet another major scientific report about the climate crisis. That unsteadiness means there is potential for many possible outcomes—but no guarantees. It will all depend on what we do and how we leverage our democratic systems to shape our collective future.

But the signs of democratic decay run deep. Naomi Klein has articulated this problem of timing: As scientists in the 1980s started to * document signs of the impending climate crisis, politics took a turn away from the collective (and against regulation), while culture deepened into consumerism.

The preferences of economic elites far outweigh the preferences of average Americans when it comes to policy outcomes in government. Historic wealth disparities and partisan polarization fracture not just families and communities but our whole nation. We are only just beginning to understand the depth at which foreign adversaries stoke these divisions with social media weaponry. Deaths from despair have risen, particularly among White, working-class Americans, who are dying earlier from suicide, drug overdose, and alcohol abuse. Across the country, Black and Indigenous people and other people of color face stark disparities in opportunity and well-being, with fissures in justice along lines of race.

It's no wonder people are angry.

And maybe it's no wonder that young people today believe less in the necessity of democracy than their older counterparts. A recent * study found that 30 percent of people born in the 1980s believe it is "essential" to live in a democracy, versus 72 percent of people born in the 1930s.

In the face of public anger and apathy, and democratic decline, the need for desirable transformation goes much deeper than climate. We need institutional and cultural tools and spaces that enable everyone to contribute. In other words, the work of our time is to reweave the fabric of our democracy even as the broken systems seem overwhelming, even as the work we each can do feels so small. It's time for an intentional practice of climate citizenship.

The word "citizenship," of course, is fraught in today's political climate. Some factions use the legal status of citizenship—and, in particular, the lack of it—in horrifying ways that sow fear, mistrust, and civil and human rights abuses. Wariness exists about the words "citizen" and "citizenship," and for good reasons. But to cede the idea of citizenship to undemocratic forces is something I am unwilling to do. It remains a powerful description for ways of being and acting in public life, upon which democracy depends.

Citizenship, at its core, is a sacred trust between the individual and collective. As we face the climate crisis, this trust—and how we understand and act on it—is more critical than ever.

Only through the collective can an individual enjoy goods such as a healthy environment and certain kinds of support and security. At the same time, as a citizen, the individual has both the right and the responsibility to hold the collective accountable and to participate in it. In other words, citizenship is a dynamic process of consent and dissent of individuals as part of a larger whole. It's how an individual can most fully satisfy fundamental human desires to be part of a community and to have agency in determining a shared future.

As a citizen, I define myself as part of a local community, a state, a nation, and the world. In my experience, only by making these identities part of who I am can I live the life I aspire to. It's a life of knowing my neighbors well, understanding the politics and governance of my place, and working with both for our future.

When I claim and allow myself to be claimed by citizenship, I declare that I am inextricably part of my community. Therefore, I'm not okay with huge disparities in wealth, health, and income, especially when they fall along lines of race. I'm not okay with economic elites having undue sway over policy outcomes. I'm not okay with fossil fuel use barreling forward or the climate vulnerability so many people already face. Because I am a citizen of this community, because I belong, I feel responsible for reshaping it into a place where we all belong better.

As part of Earth's community, we are also called to responsibility in and for it. That is the meaning of climate citizenship.

Abolitionists, labor organizers, suffragettes, civil rights activists, and feminists have shown us that the work toward genuine democracy in this country has never been easy. But the climate crisis brings new challenges to securing human dignity. For example, climate impacts are already forcing the movement of people around the planet and across national borders, accelerating migration pressures. As governments "protect" their borders, the treatment of people trying to cross them can be cruel and transgress fundamental human rights. Climate citizenship involves taking on the work of ensuring that mass human migration does not lead to mass human rights abuses. It means that we are planetary citizens in addition to being citizens of nations.

After serving six years in state government, I decided not to run for reelection in 2012. Instead, I focused on building a nonpolitical career, growing my family, and finishing a PhD. But another turning point appeared on my path of understanding climate citizenship, this one more subtle.

In 2018 my family and I spent the summer in Brno, Czech Republic, where my husband's company has an office. The three of us—my husband, our toddler, and I—set off with a sense of grand adventure. As we settled into Brno, my daughter and I developed a pattern in our days. When my husband left for work in the morning, she and I headed for one of the many playgrounds within walking distance. We'd spend a couple of hours puttering around; then we'd head back to the apartment, stopping to buy whatever was in season at the centuries-old produce market around the corner.

Our pattern worked well until mid-July, which is when the heat came and stayed for the next six weeks. That heat grounds my memories of Brno. By 8:30 A.M., I dripped in sweat as I pushed the stroller down the sidewalk. I quickly learned which playgrounds had sun protection and which didn't. I adapted to sleeping without air-conditioning or blankets but started to dream of wearing a long-sleeve shirt.

In mid-August we planned a trip to Prague, figuring we couldn't spend a summer in the Czech Republic without seeing this historic capital city. A few days before our planned departure, my daughter came down with a terrible stomach virus, and the weather predictions were for 95 degrees Fahrenheit all weekend. My husband and I didn't think we could keep our still-recovering daughter safe in that heat, so he urged me to go on my own.

I tromped all over the city, guzzling water and gushing sweat. One afternoon I wandered into the Museum of Communism, and with no other demands on my time, I read every single placard. Details of the Czech people in the twentieth century leaped out. They were buffeted by fascism and then oppressed by communism. And yet they managed to hold tight to a sense of dignity and to assert their claim to democracy and self-governance, ultimately resulting in the Velvet Revolution in 1989. That afternoon, a year and a half into the Trump presidency, I stood across an ocean from my own country and finally let myself really feel something I had suppressed—the fragility of democracy, a deep, buzzing uncertainty about its future in America. Then I walked out into the sauna-like city and felt the weight of an uncertain, unstable climate future on my skin.

After I returned home from Brno, my dissertation consumed my time and mind. I didn't immediately process what I encountered that summer—climate-change-charged heat and democracy's fragility—but those strands continued to move through my thinking for months.

Living in a democracy is not a given, though I've taken that for granted for more than three decades. Just the opposite, in fact. Democracy is quite rare over human history. And living on a planet with a climate conducive to our flourishing as a species? That is rare beyond measure in the sweep of the planet's past.

Even though the threads of climate citizenship run through my life, only in confronting the fragility of *both* democracy and this planet did I uncover its full necessity and potential.

Just as I learned from voters in 2006, my education today comes from witnessing and working alongside people who are trying to reweave democracy from the ground up and secure a safer climate through their work—from the local to the global. These folks show up at community meetings, even when those meetings are poorly run. They actively inquire how a given building proposal will deal with the increase in storm water expected in an ever-warmer world. They send letters to the editor and visit their elected officials to advocate on ambitious climate policy. They knock on doors for candidates with strong climate plans and then hold those folks accountable after electing them. They work hard to turn out the 10 million environmentalists who were registered to vote in the 2016 presidential election but didn't show up to the polls.

Some climate citizens master the tedious public processes of land-use development and then get themselves appointed to the city planning commission to be advocates there. Others become experts on the intricacies of public utility commissions—government entities that often decide the fate of fossil fuel infrastructure. There are climate citizens who organize church congregations to go solar and then invite elected officials to see the rooftop array. Public libraries are hosting community conversations on climate change. Young people around the world are ratcheting up demands for climate action at the level commensurate with the crisis, and their voices are being heard at unprecedented levels.

It's exhilarating to see people find their unique contributions to this big collective effort, and it gives me courage to forge ahead amid uncertainty. I believe that my actions as a citizen matter for shaping our climate future. And I believe yours do too.

Together, we are a climate citizenry. We wade collectively through the paralysis of fear, grief, shame, and hopelessness and into action that brings feelings of strength, possibility, and even joy. For this is noble and necessary work, and it is impossible to do alone.

Dead Stars

Ada Limón

Out here, there's a bowing even the trees are doing.
Winter's icy hand at the back of all of us.
Black bark, slick yellow leaves, a kind of stillness that feels
so mute it's almost in another year.

I am a hearth of spiders these days: a nest of trying.

We point out the stars that make Orion as we take out
the trash, the rolling containers a song of suburban thunder.

It's almost romantic as we adjust the waxy blue
recycling bin until you say, *Man, we should really learn
some new constellations.*

And it's true. We keep forgetting about Antlia, Centaurus,
Draco, Lacerta, Hydra, Lyra, Lynx.

But mostly we're forgetting we're dead stars too, my mouth is full
of dust and I wish to reclaim the rising—

to lean in the spotlight of streetlight with you, toward
what's larger within us, toward how we were born.

Look, we are not unspectacular things.
We've come this far, survived this much. What

would happen if we decided to survive more? To love harder?

What if we stood up with our synapses and flesh and said, *No.
No*, to the rising tides.

Stood for the many mute mouths of the sea, of the land?

What would happen if we used our bodies to bargain

for the safety of others, for earth,
            if we declared a clean night, if we stopped being terrified,

if we launched our demands into the sky, made ourselves so big
people could point to us with the arrows they make in their minds,

rolling their trash bins out, after all of this is over?

Wakanda Doesn't Have Suburbs

KENDRA PIERRE-LOUIS

I don't know when I first learned that the story of humanity was one of profound wrongness. But the idea that the world was broken because of our presence was always there, a constant cadence with which to orient my life, as seemingly natural as the four seasons that ordered my years and the cycle of the sun that ordered my days.

I grew up in the Catholic Church, whose theology of original sin means that humans are born with a built-in desire to disobey God ever since Eve, allegedly, took a bite of that damn apple. Since her and Adam's punishment was to be kicked out of the Garden, it's not a stretch to say that I also grew up with a belief that humans have an innate tendency to destroy their environment.

Growing up in New York City, secular classes taught to state standards imparted a similar message of humanity's inherent shortcomings. A critical undercurrent of history and science classes, for example, is the idea of perpetual progress. That progress, we are taught, has always come at the cost of the environment—from smog-filled skies to landscapes devoid of birds, as chronicled in Rachel Carson's *Silent Spring*. And, of course, the idea of progress itself implies that there is something wrong with the present and with our place in it.

I am not the only person who has gotten this message. Robin Wall Kimmerer, an enrolled member of the Citizen Potawatomi Nation, scientist, author, and lecturer, writes in her beautiful book *Braiding Sweetgrass* that in her survey of roughly two hundred third-year ecology students, almost all said that they thought that humans and nature were a bad mix. These were, she pointed out, people who had chosen to devote their lives to environmental protection.

"I was stunned," she wrote. "How is it possible that in twenty years of education that they cannot think of any beneficial relationships between people and the environment?"

When I share this quote with a friend, she points out that it is ahistorical to frame people and nature as always in opposition with each other. There are societies that live within their ecological boundaries.

The mountainous nation of Bhutan, for example, absorbs more carbon than it emits. But good luck finding those stories in the tales that dominate much of U.S. culture.

As someone who consumes what I'm told is an excessive amount of popular culture, I can rattle off entire genres of film—from dystopian science fiction to horror movies to serious dramas—that reaffirm the idea that where humans go, ecological devastation inevitably follows.

Movies like *Mad Max Beyond Thunderdome*, *Waterworld*, and *The Hunger Games*—all of which, to be clear, I enjoyed—posit not only that humans could decimate the planet but that we will. Arguably my favorite dystopian movie, the 1986 *Solarbabies*, which features a glowing alien orb and kids roller-skating through the desert, is based on the idea that a global corporation would happily lock up every drop of water to turn the planet into a giant desert, and by proxy enslave humans, out of a desire for world domination. Reflect on that for a moment: The film's central conceit is that lots of people think turning Earth into a hellscape is preferable to sharing resources.

James Cameron's popular movie *Avatar* features a planet where intelligent beings live in concert with their environment until humans, once again, show up to destroy it. The short-lived television show *Terra Nova* deals with an overly polluted, overpopulated Earth by sending humans back in time to when dinosaurs roamed—before humans had a chance to spoil the place. The youth-oriented television network the CW has a show called *The 100* in which the Earth is "simmering in radiation" as a result of a nuclear war. In the series opener, the surviving humans orbit the now-hazardous planet on spaceships. Space also features prominently in *WALL-E*, an animated film by Pixar that is ostensibly a story for children. It features not just a polluted Earth but an atmosphere so physically littered with trash that humans have opted to live on sterile space stations. Even *Star Trek*, which many hold up as a positive vision of humanity's future, assumes that in order for us to live in equilibrium with our environment, we must first let key species, including whales, go extinct; fight a third world war; have a wiser, more advanced alien species intervene in our development; and create an off-planet resource base.

No big deal.

So disabused are we of the notion that humans en masse can have anything approaching the good life, that some of the most terrifying pieces of fiction aren't ones in which the opening scenes are filled with horrors out of Dante's *Inferno* but rather ones in which the setting is bucolic. In the television show *The Good Place*, demons torment the show's protagonists by wielding a slightly twisted version of heaven. From Shirley Jackson's short story "The Lottery" to Jordan Peele's film *Get Out*, idyllic settings are increasingly a signal to the audience to prepare for some underlying wickedness.

Of course, these are "only" stories. But stories are powerful. A 2017 study in the journal Nature Communications found that among hunter-gatherer societies, those with better storytellers are more cooperative. It's our ability to cooperate that anthropologists say has allowed humans to survive even the harshest environmental conditions and to fend off predators that could take us out individually. In fact, our brains are wired for story, which is why most of us can retain stories far longer than we can retain a series of facts stated plainly. The stories that we tell about ourselves and our place in the world are the raw materials from which we build our existence. Or, to borrow from the storyteller Kurt Vonnegut, "we are what we pretend to be, so we must be very careful what we pretend to be."

Simon Lake, who helped pioneer the modern submarine, admits he was influenced by the idea of undersea exploration after reading Jules Verne's *Twenty Thousand Leagues Under the Sea*. Physicist Leo Szilard, who helped develop the atomic bomb, was motivated in part by the H. G. Wells novel *The World Set Free*. The *Star Trek* communicator was the inspiration behind the cellphone.

Increasingly, the idea that the Earth will become unlivable due to human actions has created a push for us to move to a new frontier. It's unclear if tech billionaires are taking cues from *WALL-E*, but some, such as Elon Musk and Jeff Bezos, are investing heavily in space exploration, their efforts predicated on the idea that humans need to move into space to survive as a species.

"We humans have to go to space if we are going to continue to have a thriving civilization," Bezos told *CBS Evening News* in 2019, envisioning a world where we put big factories in space and zone the Earth as residential.

That humans would head out to space in order to mine it, by the way, is the framing of the television show *The Expanse*, which was saved from cancellation by Amazon, of which Bezos is founder and CEO. Like me, Bezos is on record as being a fan of the show.

Not everyone agrees with his must-go-to-space story, however.

"Fun fact: space wants you dead. Mars wants you dead. Titan wants you dead. Space will fucking kill you," tweeted Shannon Stirone, a science writer whose work has appeared in *The New York Times* and *The Atlantic*.

"Earth is Easy Mode," tweeted Mika McKinnon, a Canadian field geophysicist and science communicator, in agreement. "If we can't maintain habitability here, we're utterly fucked trying to pull off long-term survival anywhere else."

And yet billions of dollars are being allocated to this effort. What does it say that spinning a story about humans moving into a radioactive vacuum resonates more strongly with many people than our chances of reducing greenhouse gas emissions?

Right now, the stories that many of us are telling about ourselves are hurting us.

It is not universal—there are cultures Indigenous to North America, for example, that are still enacting a different story—but this story of inherently destructive humans is the most mainstream. We need different stories, ones that help us envision a present in which humans live in concert with our environment. One in which we eat, play, move, and live in ways that are not just lighter on the Earth but also nurturing to us as humans, with at least some of the trappings that many of us have come to expect of modern life.

In fact, I can recall only one mainstream vision of the present—never mind the future—that starts from the premise that humans can live in equilibrium with their environment.

Our first glimpse of the fictional African country of Wakanda—the backdrop of the Marvel movie *Black Panther*, which is loosely based on the Mutapa Empire of fifteenth-century Zimbabwe—begins with the nation's natural landscapes. Wakanda is the most technologically advanced country in the world, but the filmmakers choose to ground our understanding of this civilization in sweeping images of its mist-veiled mountains; its verdant valleys, where sheepherders drive their flocks in ways they have likely done for generations; and its border tribesmen, wrapped in the Wakandan version of Lesotho blankets, galloping by on horseback. The message seems to be that Wakanda is a country whose greatest technological achievement is maintaining its environment.

From there, the camera takes us on an aerial tour of the capital city, Birnin Zana. The transition from lush countryside to bustling city is so abrupt, so fantastical, that it's easy to miss what the camera is really telling us: that Wakanda can maintain its ecosystems in part because there are no suburbs.

It's a vision of a modern civilization that looks and feels drastically different from the United States, where an estimated 52 percent of Americans describe their neighborhoods as suburban. Suburbs emerged in part because cities were polluted, with bad air quality and substandard housing. But rather than address those issues, we simply relocated our population and created new issues, not least among them an increased climate impact. Studies have shown that suburb dwellers have greater greenhouse gas emissions than their urban counterparts.

Black Panther's vision of Wakanda rejects the oft-repeated story that we humans and our environment are natural enemies. Instead, it tells a story in which humans have become technologically sophisticated while maintaining a flourishing relationship with their surrounding environment.

To paint this picture, the filmmakers borrowed heavily from existing African cultures—the Maasai people of eastern Africa, the Zulu people of South Africa, the Sotho people of southern Africa, and the Himba tribes of Namibia, among others—and asked themselves: *If these groups had not been colonized and had opted to live within their ecological limits, what would the resultant society look like?* And to them, it looked like an urban core surrounded by countryside before giving way to true wildlands.

In Birnin Zana the skyscrapers, as envisioned by director Ryan Coogler, are high, organic structures that rise out of surrounding forests so lush they have likely never been chopped down. And many of the buildings have plants growing off terraces and rooftops. On the ground, a trolley toddles through the city, kicking up some dust because the ground is not asphalt but rather pressed earth. It moves fast enough to beat walking but slowly enough that people instinctively move out of its way, mingling through a narrow (by North American standards) streetscape that contains no sidewalk—evoking the famed words of the sociologist Lewis Mumford, "Forget the damned motor car and build cities for lovers and friends." Wakanda doesn't have cars, and thus its streets have no need for separation.

All of this means that even without vibranium, the fictional metal that powers Wakanda's technology, the country would be more sustainable than the United States. Cities that have more green space are naturally cooler, requiring less air-conditioning, lowering energy use and making it easier to reduce fossil fuel use and rely on renewable energy. They also reduce the likelihood of minor flooding because, unlike cement or concrete hardscaping, soil absorbs water instead of having it run off. Green spaces also provide a host of psychosocial * benefits, from improving mood and behavior to speeding up healing and increasing students' ability to learn.

Eliminating suburbs and their longer-distance counterpart, exurbs, could be one way of slashing the country's ecological footprint. No suburbs mean fewer roads to fragment ecosystems, making it easier for animals and plants to survive, promoting biodiversity. Roads also use a lot of cement, which is a major source of greenhouse gas emissions, as well as sand, which is often acquired by dredging rivers and other critical ecosystems. These impacts happen regardless of whether or not the vehicle in question runs on gas, electricity, or the hopes and dreams of future generations. It's the whole structure of suburban life and car dependence that's the problem.

At the same time, it's not clear that suburban living, which is associated with increased social isolation, especially for those with longer commutes, is any good for humans. At least one study, of couples in * Sweden, found that those where at least one partner has a commute forty-five minutes or longer are 40 percent more likely to divorce. Suburban residents are less likely to get physical exercise than their urban counterparts.

Beyond its fantastical elements—superpowers and impossibly high tech—*Black Panther* is telling a very different story of what it means to be human. And even when trouble comes, it comes in the form of an angry outsider named Killmonger, who is telling a flawed story based on conquest and scarcity—a very U.S. story—of what it means to be human.

Wakandans elected to tell a story about themselves that differs from Killmonger's—namely that it was possible to improve the quality of their lives without degrading the environment that they depend on—and then they did it.

And yet when we talk about climate change, there's often a hidden resignation—like, of course we harmed the Earth. And when we talk about acting on it, there's also an undercurrent: that it will require a level of sacrifice that is worth it, but just barely. What if, instead, the story we tell about climate change is that it is an opportunity? One for humans to repair our relationship with the Earth and reenvision our societies in ways that are not just in keeping with our ecosystems but also make our lives better?

If this doesn't sound possible, ask yourself: Why not?

Maybe we should start telling ourselves a different story. One that is a bit more like Wakanda. One that maybe goes something like this . . .

Once upon a time, some humans told a story about their relationship to the Earth, and they used it to build a world that was beautiful but flawed. Over time, people realized that was the wrong story and they constructed a new one, one that said they could live in harmony with their environment. And they used the pieces of their old story to help construct their new one.

4. RESHAPE

ENOUGH OF DISRUPTION AND EXTRACTION—

HOW BEAUTIFUL, THE QUIET WORK OF MAINTENANCE:

TO REPAIR THE WORN-DOWN,

TO MEND THE TORN

MADELEINE JUBILEE SAITO

Heaven or High Water

Sarah Miller

"Sunny day flooding" happens when water comes up through the ground. It's caused by the tides, even on days when there is not a cloud in the sky, hence the name, but it can certainly also rain during sunny day flooding, and yes, that makes it worse. This phenomenon has become a regular occurrence in places like Norfolk, Virginia, and Charleston, South Carolina. Sunny day flooding happens in many parts of Miami, Florida, but it is especially bad in the low-lying area of Miami Beach—the wealthy island separated from the city proper by Biscayne Bay.

The sea level in Miami has risen ten inches since 1900. In the two thousand years prior, it did not really change. The scientific consensus is that the sea will rise in Miami Beach somewhere between fourteen and thirty-four inches above 1992 levels by 2060. By 2100 it could be six feet, which means, unless you own a yacht and a helicopter, sayonara. Amazingly, in the face of these incontrovertible facts about the climate, the business of luxury real estate is chugging along just fine, and I wanted to see the cognitive dissonance up close.

Lying is not my favorite, but when it's called for, the only thing to do is jump in with both feet.

When the first real estate agent—tall, fair, polite bordering on stern—possibly Swiss, possibly Swedish—asked, "Do you live in Miami now? Do you know what kind of place you're looking to buy?" I said, "I live in San Francisco and my husband is in tech." I gave a coy twist to the wedding ring I'd put on in my hotel room. "We're looking for . . . a place to hang out when it gets really rainy [lol] and then to retire to [roflmao]."

He either believed me or did not give a shit.

The decor was beige and white with stainless steel, except for the books on the nightstand, which were jewel-toned. I walked around the condo as if I already owned it, as if within my lifetime the lobby beneath us would not be decorated with kelp.

The realtor and I rendezvoused on the balcony, which overlooked

Biscayne Bay. He gestured at the rainy day, unusual for this time of year, late March. "Usually at night, you will be looking at the best spectacle of a sunset here," he said. I oohed and aahed over the view, quite genuinely, because if you don't think about the fact that it's filled with thousands of pounds of post–hot Pilates ceviche poops, Biscayne Bay is breathtaking.

I asked how the flooding was.

"There are pump stations everywhere, and the roads were raised," he said. "So that's all been fixed."

"Fixed," I said. "Wow. Amazing."

I asked how the hurricanes were.

He said that because the hurricanes came from the tropics, from the south, and this was the west side of Miami Beach, they were not that bad in this neighborhood. "Oh, right," I said, as if that made any sense.

I asked him if he liked it here. "I love it," he said. "It is one of the most thriving cities in the country; it's growing rapidly." He pointed to a row of buildings in a neighborhood called Edgewater. "That skyline was all built in the last three years."

"Wow," I said, "just in the last three years. . . . They're not worried about sea level rise?"

"It's definitely something the city is trying to combat. They are fighting it, by raising everything. But so far, it hasn't been an issue."

I couldn't wait to steal this line, slightly altered: *I am afraid of dying, sure, but so far, it hasn't been an issue.*

Later I texted Dr. Kristina Hill, an associate professor of urban design at the University of California at Berkeley, whose main work is helping coastal communities adapt to climate change. I told her the agent's theory about weaker hurricanes on the west side of Miami Beach. She wrote back, "That's ridiculous!"

The next open house was not far. I popped into a store en route, in an area where the sidewalk had been raised. "There used to be flooding here," the owner said as she folded a soft sweater. She had long, dark hair and, as was de rigueur in Miami Beach, lash extensions. "But they put in pumps and it's been fixed."

"So I hear," I said.

"Yeah," she said. "It's amazing."

"I don't know if I understand this," I said. "The sidewalk is raised, but where does the water go?"

"Into the drain," she said. "Well. Except for one time. One time the store was flooded. But it's fixed." She put her hand over her heart in an expression of extreme gratitude.

The next real estate agent was in her midforties and wore a diamond ring big enough to plate a filet mignon. Her pants were so perfectly tailored she looked like she'd been sewn into them.

I walked around this property in the slow, pensive way of the rich shopper, cultivating an opaque expression which could suggest equally the taking in of beauty or polite condemnation. The place was lovely, with a water view and, like everything in Miami, beige, beige, beige, pink, white, beige, blue, beige, beige, white.

The rhythm of these things is as follows: greeting, walk around, short chat, goodbye. This short chat was longer: We talked about shoes and jewelry and the intense beauty of Miami, which I meant every word of. I felt bad lying to her, and with no good segue for my true mission, I was worried that when I came out with my questions, her demeanor would change. But just as charmingly as she had received my greetings and compliments on the layout of the kitchen, she said sure, there was a problem, but if anything was going to happen, she thought it would be more like in fifty years than in thirty.

It's amazing that people in these situations tell you what they *think*. *I think bread actually takes twenty minutes to bake*, she said, removing the doughy mass from the oven. *I think I can drive a car after I've run out of gas*, he said as he rolled silently into the breakdown lane.

The agent continued: "The scientists, economists, and environmentalists that are saying this stuff; they don't realize what a wealthy area this is." She said that she lived here and wasn't leaving, and that the people selling Miami were confident and all working as a community on the same goal, to maintain this place, with the pumps and the zoning and the raising of streets. There were just too many millionaires and billionaires here for disaster on a great scale to be allowed to take place.

"Anyway, people are working really hard to prevent anything from happening," she continued.

Another agent came in to look at the condo and joined our conversation. She was young, and if indeed we are talking thirty years until "Miami Beach–pocalypse," the forty-something realtor and I will very possibly be dead, or close to it, when shit really hits the fan, but this woman will still have many good years left. Still, she did not seem to be losing a great deal of sleep over sunny day flooding, sea level rise, any of it.

"From what I understand," she said as she took a turn in the condo, her heels clacking across the pale floors, "everybody has done this, like, research, and they have these, like, like . . ." She paused behind the kitchen island, her pastel nails splayed out on the varnished counter top. "I can't think of the word now."

"Studies?" said the other realtor helpfully.

"Yeah," replied the younger woman. She said she knew about a guy who had "paid for, like, a study. And basically it said we shouldn't be concerned because it's being figured out. Unless you have a family, and you're planning on staying here."

The ideal buyer for this place was someone who was okay with the street and lobby being full of water for the next couple decades, at which point they might actually have to leave, unless it all got "figured out." And apparently those people exist, lots of them. "A lot of people just buy something here, they keep it for five years, and then they sell it," she explained.

I had a bit of trouble finding the next property. I felt like an idiot until several phone calls determined that the place did not exist yet. When I got to the showroom, the staff were all in a mild state of amused agitation because a news crew had arrived to speak to the famous architect of this building-to-be and there had been a water-main break in front of it.

The woman showing the model unit wore rubber boots with her fancy real-estate-lady outfit. The building was so luxurious—literally every inch of it a visual spectacle, with polished fixtures, views around every corner, wide vistas of plush upholstery—that it was difficult to concentrate on what she was saying.

I did manage to tune in when she told me that the building had been built several feet above the zoning requirements. This was a big deal as a selling point, it seemed. I thought about a line from Jeff Goodell's book *The Water Will Come*: "There is always the risk that Mother Nature won't respect the design specifications."

I said it was surprising that such a famous architect would build in a city so threatened by climate change. I didn't actually find this surprising at all, since everyone needs to make money, but I wanted to see what she'd say.

She said the main thing was just that Miami was being very forward-thinking. She mentioned Amsterdam and how they were *making it work*, and how the Dutch were the poster child for sorting out a way to *make this work*. "I think the takeaway is just that Miami is doing something about it."

There are several problems with comparing Miami to the Netherlands. One is that the Dutch have spent billions of dollars on climate resilience and Florida has spent millions. The Dutch strategy is holistic—accounting for how this thing will affect that thing, etc.—whereas in Miami they have just installed some pumps and raised some roads and buildings, which kind of ignores the issue that a place to live is really only useful insofar as nearby goods and services, and the streets and sidewalks for reaching them, are not underwater.

As I listened to these people's reassurances that everything was fine, I started thinking that maybe I was crazy. I spoke to Dr. Astrid Caldas, a senior climate scientist at the Union of Concerned Scientists. According to their projections, by 2030 there will be about 45 days of sunny day flooding per year in Miami. By 2045 there will be about 240—in other words, two out of every three days. She confirmed my suspicion that while the raising of buildings was good for the buildings, it didn't do much for the well-being of those living inside. "Yes, you do need to be able to get out of the building to get medicine and groceries," she said. "If all the streets are flooded, what then?"

With all the talk about pumps, I had expected something truly enormous. But they were the size of maybe the world's largest Airedale. When I mentioned them to Dr. Hill, she gave a skeptical snort. "Yeah, in Miami, those pumps and those raised roads—that's, like, their big move," she said. "But it's just kind of cosmetic." She acknowledged they help with flooding. But what about when the sea begins to rise significantly, or when there's a big storm event? A big storm would not just flood everything and cause damage and then retreat. It could unearth septic systems, which could lead to terrible disease, or industrial waste, which could poison people. "That's the situation that really concerns me and is, as I see it, the canary in the coal mine," Hill said.

I talked to Dr. Amy Clement, a professor at the University of Miami's Rosenstiel School of Marine and Atmospheric Science, about pumps and raised buildings. "No, you're not crazy," she reassured me. "That alone is not coordinated planning, and it's not a comprehensive solution." She told me about a legal battle between homeowners and the county government in St. Johns County, near Jacksonville, Florida. The homeowners said the county was depriving them of access to their land; the county said it would no longer foot the bill to maintain a road continually ravaged by storms and erosion. "People are just assuming the government will maintain their roads, and that may not always be the case," Clement said.

Then there is the problem of walls. The big plan in the Netherlands depends on walls. Since Miami is built on limestone, which soaks up water like a sponge, walls are not very useful. In Miami, seawater will just go under a wall, like a salty ghost. It will come up through the pipes and seep up around the manholes. It will soak into the sand and find its way into caves and get under the water table and push the groundwater up. So while walls might keep the clogs of Holland dry, they cannot offer similar protection to the stilettos of Miami Beach.

Miami Beach is not the only threatened part of Miami. There are plenty of neighborhoods, like Shorecrest or Hialeah, where flooding is also bad. But while wealthy Miami Beach is fussed over, every scrap of attention or money lower-income areas receive they must beg for.

People say Miami is douchey, but really, I loved almost everything about it: the symmetry of the blue umbrellas on the beach, riding a bike under a canopy of trees, sitting on a wall watching the sunset, definitely not thinking about how seawater might be infiltrating the septic systems behind me. The whole time I was there I was like, *Yeah, I can see why no one wants to admit how fucked this place is.*

That night I went out to dinner with a friend who grew up in Miami and had left for college twenty years before, never expecting to return. He was in elementary school when Hurricane Andrew hit, and that was when he realized Miami was not going to last forever.

He moved back in 2018, after years away, and saw that the party was still on, even though perhaps it shouldn't have been. That said, it was perhaps on for this night, for there we were at NIU Kitchen, downtown, drinking a really good wine from the Languedoc, surrounded by extremely good-looking people, enjoying luxury while discussing the horrors that luxury has visited upon the world.

My friend is active in the local civic community but says he's skeptical even of the activist discourse around sea level rise. "There's all this talk about 'sustainability' and 'resilience,'" he said, "and it kind of sounds to me like 'What's the least we can do in order to keep the party going?'"

I told him about someone I knew who had gone to a meeting about climate change where Miami officials talked about how they had to demonstrate to the world that they were all about resilience, and how she had been amazed that they thought that was actually the extent of their job, to just convince people they were on top of things when they absolutely were not.

Get more efficient and find the right incentives to encourage the right kinds of enterprise—that's the neoliberal thinking on the "reasonable" way to approach this stuff. But, my friend wondered, *what if the mature thing to do is to mourn—and then retreat?*

The next day was beautiful; the streets were dry under a blue sky. It was a great day to think Miami would last forever. The first property I looked at, on the ocean side of Miami Beach, was asking $4.1 million.

"Frankly, this is a little much for me," I said to the agent, "but I'm just getting a sense of the neighborhood." If a young Robert Redford ever fantasized about giving up a few degrees of handsomeness just to be tall, it was this man that he pictured.

"Happy to see West Coast coming east," he said.

I walked upstairs. I did my thing, which was to take in the splendor of concrete sinks and the guest soaps, each delicately adorned with a sprig of greenery. Then there was the round of compliments about the space, the small talk, and the sea level question put forth with my guileless naïveté.

"You know, they're picking on Miami because we don't have our heads in the sand," Robert Redford's tall cousin said. "We're actually being proactive about it." He pointed out to me that as I arrived, I had actually stepped up several feet. "That helps a lot," he said. I nodded and did not say, *Yes, it helps the house, but houses don't have to go outside; houses don't get cholera.*

"You feel like Miami is getting more attention because it's trying harder?" I asked.

"Yeah, because all our elected officials are like, 'Yeah, we're gonna deal with this.'" He talked about "the Zika news cycle" and how that had come and gone.

"Oh," I said, "so you think sea level rise is kind of like Zika?"

He made a face indicating he understood that it wasn't. Then he told me how they were raising everything. He told me I could get a great one-bedroom condo over in Sunset Harbour (perhaps the most vulnerable part of Miami Beach) for $900,000, that they were planning a lot of new construction there. Then he talked about the pumps, that they were hitting all the problem areas first, but then they were going to do the whole city. "They raised the streets and it fixed the problem. It probably gives us another fifty years. It's all being taken care of. And Miami is really taking off," he said, making an airplane gesture with his hand.

About that I believed him. You could not look up in this town without seeing a construction crane.

I have to admit, I kind of liked this guy. I liked all of them. They were a likable bunch. That (and being White) is how they'd gotten jobs on the front lines of capitalist hypocrisy. And we all had jobs in it somewhere. How else were we supposed to live? Of course, they made more money than most people, way more. But still, like most of us, if they didn't act like this was all totally fine, they wouldn't eat. It seems to me that as long as we live under capitalism, where the motive is profit, we are going to keep burning cheap fossil fuels and sea levels are going to continue to rise. And people have to live somewhere. I just bought a home in a town in rural Northern California that a forest fire could level any day. I'm not smarter or better than the person who will buy a home in Miami.

I rented a Citi Bike to ride to the next property. In this part of town, just west of the beach, a lot of the streets were shaded under a canopy of the most beautiful trees, with big, oval-shaped, glossy leaves.

My last real estate agent looked like Botticelli's Venus, I shit you not. Again with the ten-million-dollar slacks, tailored to perfection.

This time I didn't bother working up to my questions: "So, I mean, like, even with sea level rise, even with a thirty-year mortgage, you'd still be fine?"

"Of course!" said this beautiful woman.

"I mean, if you had an eighty-year mortgage . . . ," I ventured, trying to get into the "fuck eighty years from now" spirit of things. We laughed at the hilarity of an eighty-year mortgage.

"I mean, it's not like you're going to wake up one day and the ocean is outside your window!" We laughed again. "If something is gonna happen, it's like, I think it's like, a hundred, two hundred years."

This is what she *thought*.

She showed me the upstairs. We admired the large closets. She said she had a lot of family in Italy and began to talk about Venice, where she had been many times. "At high tide the water from the sea comes into the city, right into the Piazza San Marco. And it's horrible." Her enormous green eyes widened, reflecting the horribleness.

"They are going to have to get something for the people to walk on, for the tourists. They're going to have to put something so the people can walk on top. But every year they say the same thing about Venice, that it's going to go down." She made a face like *How do those idiots say this?*

The bathroom tiles were the color of Biscayne Bay. I said so.

"Yes!" she said. Her eyes were full of real, deep love for blueness. "Beautiful, no?"

As we walked down the stairs to the first floor, she turned to look at me. She was very earnest, standing very close. I felt her beauty soak into me. "It's Miami," she said. "We are surrounded by water! There's not a solution. But nothing is going to happen."

Man on the TV Say

PATRICIA SMITH

Go. He say it simple, gray eyes straight on and watered,
he say it in that machine throat they got.
On the wall behind him, there's a moving picture
of the sky dripping something worse than rain.
Go, he say. Pick up y'all black asses and run.
Leave your house with its splinters and pocked roof,
leave the pork chops drifting in grease and onion,
leave the whining dog, your one good watch,
that purple church hat, the mirrors.
Go. Uh-huh. Like our bodies got wheels and gas,
like at the end of that running there's an open door
with dry and song inside. He act like we supposed
to wrap ourselves in picture frames, shadow boxes,
and bathroom rugs, then walk the freeway, racing
the water. *Get on out*. Can't he see that our bodies
are just our bodies, tied to what we know?
Go. So we'll go. Cause the man say it strong now,
mad like God pointing the way outta Paradise.
Even he got to know our favorite ritual is root,
and that none of us done ever known a horizon,
especially one that cools our dumb running,
whispering urge and constant: *This way. Over here*.

A Tale of Three Cities

Jainey K. Bavishi

It was a late-December afternoon in New York City. I was in labor and exhausted. It seemed like it had been raining most of the night, but it was hard to tell from my hospital room. When I finally saw my doctor, she told me there were flood warnings across the area. My mind flashed to one of the lowest-lying communities in the city, which often floods from high tides even on sunny days. Surely some streets there would be impassable. Later I would look on the Facebook group where residents report flooding events. One resident had posted that day: "Does anyone know if [it's] safe to get out yet? Have to get to work eventually." I stared at the screen, overwhelmed: We still have so much work to do to adapt to a changing climate and adjust to the new realities it presents.

New Orleans: Whose City?

When Hurricane Katrina struck and the levees protecting New Orleans failed, in August 2005, around 80 percent of the city was flooded and more than one thousand people lost their lives statewide. Vast areas of many neighborhoods were inundated, submerged under more than ten feet of water. The storm displaced more than one million people in the Gulf Coast region. While some returned home within days, up to 600,000 households were still displaced a month later. The population of New Orleans fell by more than half, from more than 480,000 right before Katrina to about 230,000 in 2006. In 2018 there were 92,245 fewer African Americans living in New Orleans than before Katrina, compared with 8,631 fewer Whites.

Ironically, the same systems that were built to keep the city dry have led to its growing vulnerability to flooding. New Orleans was originally settled along the "sliver by the river"—a strip of naturally high ground by the Mississippi. As time went on, engineers developed complex drainage systems and flood-protection levees, allowing the city to expand into former marshland. Pumping water out of New Orleans has accelerated subsidence of the land, and levees have prevented the Mississippi River from delivering replenishing sediments to surrounding coastal wetlands that act as a buffer to storms. Due to land sinking, sea levels rising, and wetland loss, 1,900 square miles of coastal land have disappeared since 1932, and without action another 1,806 square miles will be gone by 2060, leaving the city without its natural protection.

In January 2006, four and a half months after Katrina devastated the city, the mayor-appointed Bring New Orleans Back Commission presented its first recommendations. Many New Orleanians had not yet been able to return to see what was left of their homes and belongings. Much of the city still lay in ruins with growing layers of mold. The plan's unveiling was met with anticipation as the first step in setting an agenda for how the city would recover and rebuild. It was standing room only as residents filled the meeting hall, and many more waited outside. At the time, I was a graduate student advising philanthropists on how to support community organizing and advocacy after the storm. I had squeezed inside the room to hear the recommendations firsthand.

The commission presented a map that would become infamous: the "green dot" map. Green dots had been placed on the low-lying areas most susceptible to flooding, indicating that they would be returned to wetlands or used as parks and open space—unless the residents of those areas proved their "viability" within the next four months. With many of these communities still largely displaced, and a lack of clarity on how viability would be assessed, the implication was that long-established, predominantly African American communities would not be allowed to rebuild. The commission was proposing a moratorium on construction in these neighborhoods and a relocation of residents to higher ground.

New Orleanians reacted with anger and defiance. The green-dot map confirmed the worst fears of many displaced residents: They would lose their homes and the right to return to their neighborhoods. Within weeks, the mayor rejected the commission's recommendations, and in the end no neighborhoods were declared off limits. But the question of whether it was *safe* to live in those green-dotted places remained unanswered.

That meeting prompted the "Great Katrina Footprint Debate" over whether New Orleanians should be allowed to rebuild wherever they wanted, even if it put them in harm's way. From a purely scientific stance, the best way to reduce the city's exposure to flooding was to abandon entire neighborhoods, based on documented risk. From a purely social stance, people needed to return to their homes and communities and argued that adequate levees should be built to protect everyone. The latter prevailed, and in subsequent years the U.S. Army Corps of Engineers improved hurricane defenses in the greater New Orleans area, including strengthening a 130-mile levee system, at a cost of more than $14 billion.

Soon after the system was completed, however, the corps revealed that due to sea level rise, the levee system would likely be inadequate in protecting New Orleans and its suburbs from storm surge by as early as 2023. This fact alone should have prompted a revisiting of the great debate, but this time in a way that upheld social values while acknowledging the realities of the scientific data, rather than pitting one against the other.

After Katrina, the commission attempted to take advantage of a window of opportunity to better prepare New Orleans for future risks. But the commission's recommendation failed, as it had not sought the perspectives of historically marginalized communities in the lowest-lying areas—those who would be most impacted. New Orleans's misguided efforts offer many lessons for other cities, primarily that plans to prepare neighborhoods for climate change must be developed in an inclusive way that provides ample opportunities for engagement. It also underscores some of the hardest questions that cities will have to grapple with in the face of climate change: Where can we live and where must we leave?

Honolulu: Buying Time

The 2014 hurricane season brought two back-to-back near misses to Honolulu: Hurricanes Iselle and Julio. Residents and tourists alike were told to have two weeks of supplies on hand in case the islands' food supply chain was disrupted by the storms. Hawaii relies on imports for up to 90 percent of its food, and emergency managers warned it could take a full week before a disaster-relief operation could be initiated, due to the islands' remoteness.

At the time, I was living in Honolulu and leading an organization focused on reducing risk from disasters across the Asia-Pacific region through multisector partnerships. Although Iselle and Julio veered off and left us unscathed, they provided a strong reminder of the city's vulnerability as climate change makes storms more frequent and intense. As concerns swirled about how to prepare for the next hurricane, I began convening discussions between government officials and tourism industry leaders. One clear economic priority emerged: how we might protect Honolulu's Waikiki Beach, the state's most profitable tourist destination, from chronic flooding worsened by sea level rise.

Waikiki generates 42 percent of Hawaii's tourism revenue, bringing in $2 billion each year. The beach has had erosion problems since the late 1800s, largely because builders began developing too close to the natural shoreline. Though more than eighty structures, including seawalls and rock jetties, have been erected over time, erosion continues to claim one foot of beach every year. While tourist numbers have so far remained strong, 12 percent of visitors surveyed in 2008 said they would not come back because of the small, overcrowded beaches. And sea level could rise three feet by midcentury.

During conversations about the future of Waikiki, it became clear that saving the beach was a priority not only for the hotels but also for residents. The tricky part was figuring out how to do it. Seawalls could protect private property but would only further erode the beach. Rock jetties would protect the beach but break the waves that surfers expect and enjoy. Everyone agreed that investing in protecting Waikiki Beach was important—especially given that high tides were already flooding it—but which protections to invest in was the subject of a heated debate.

The question that caused the most discomfort was whether *any* of these measures could keep up with the rate of erosion, especially as sea levels rise. It's not clear how much time erosion-control measures can buy, and there was a certain resignation that in the long term, the challenge would become unmanageable. Everyone agreed that, for now, it made economic sense to proceed with protecting the beach in some way.

In 2015 the Honolulu City Council passed legislation requiring businesses along Waikiki to help pay for long-term beach management. In 2019 the resulting funds, combined with an investment from the state, helped to implement a large sandbag groin perpendicular to the shoreline to trap sand, addressing an erosion hot spot on the beach—temporarily.

Honolulu's predicament is not unique. While cities need to focus on financing, designing, and implementing near-term adaptation strategies, many of these important efforts simply cannot address long-term threats. Even the best-designed approaches have their limits. In places like Waikiki, we are only buying time. Even so, it's time we desperately need while the world gets its act together to achieve carbon neutrality and develop longer-term plans for comprehensive solutions. There's a finer lesson here too, which is that cities will have to continually make decisions about where and how much to invest in the face of a great deal of uncertainty. No one knows exactly how much time Waikiki has left. Honolulu officials will have to monitor how new erosion-control measures hold up, and continue to assess how long they're likely to remain effective based on the latest science about sea level rise. This points to an unsatisfying but important truth: Adaptation is a process, not an outcome.

New York City: Transformation

New York City has 520 miles of coastline, more than Miami, Boston, Los Angeles, and San Francisco combined, and 400,000 residents live in its floodplain. In 2012 Hurricane Sandy took the lives of forty-four New Yorkers, caused $19 billion in damage, and flooded nearly ninety thousand buildings.

The Lower East Side, a racially and linguistically diverse neighborhood with a median income of $40,000 and one of the highest concentrations of public housing in the country, endured major damage from the storm. Ninety-four percent of residents lost power; homes and subway tunnels flooded; and grocery stores, pharmacies, and other businesses were closed for days on end. In the wake of the storm, community-based groups served as first responders, assessing the needs of residents and distributing life-saving supplies. More than half of the neighborhood's residents reported getting assistance from a community-based organization, tenant association, or religious group in the days and weeks after Sandy—a reminder that the basic tenet of neighbors helping neighbors is critical for weathering extreme events.

Beyond the immediate aftermath, the community's engagement was vital in shaping an effort that would come to be called the East Side Coastal Resiliency Project. The project was initiated with a grant from the U.S. Department of Housing and Urban Development's Rebuild by Design competition, which brought together communities and professional designers to propose visionary and innovative solutions to prevent flooding.

At the heart of this project is East River Park: a sprawling waterfront expanse flanked by the East River and the FDR Drive highway. Sandy exposed the vulnerability of this area, which will face up to 2.5 feet of sea level rise by midcentury, along with intensifying coastal storms. Breaking ground in 2020, the project will raise East River Park roughly eight feet, creating flood protection at the water's edge. The elevated park will also connect to a series of floodwalls and floodgates stretching farther north and south along the coastline. When complete, it will protect approximately 110,000 New Yorkers—a small city unto itself. The project aims beyond building physical resilience to climate impacts, to include the social resilience of communities.

By the time I started my role as the director of the Mayor's Office of Resiliency, community engagement in this project had already been under way for four years. To be sure, there were moments in the project when city agencies could have been more transparent about behind-the-scenes technical analysis, but public input was consistently incorporated throughout. Residents raised the need for better access to the park; existing pedestrian bridges over FDR Drive are narrow, making it hard to reach the park with strollers and wheelchairs. The project design includes new pedestrian bridges to create more inviting, inclusive entry into the park. Through community meetings, we heard that the park should be used for more than Little

League and other sports; residents sought space between fenced-off fields and courts to gather, have picnics, and enjoy waterfront views. Designers in turn maximized the park's open, interstitial spaces. The design would have been very different (and, frankly, not as good) without rich community input.

This project is an example of the opportunity that adaptation investments present. It will transform the waterfront of the Lower East Side with a park that will serve the community while also protecting it. Achieving multiple benefits is an adaptation planner's dream, but it's not easy. Every neighborhood is different—the geography and topography, the available land, the density of infrastructure, the waterfront usage all vary. Implementing resiliency measures in a place as dense and constrained as New York City is extremely complex. Solutions must be tailored to the unique conditions of each place. While the East Side Coastal Resiliency Project may not be easily replicated, the principle it represents is an important one: As cities prepare for climate change, we have the opportunity to build more vibrant communities.

The Future of Coastal Cities

Our coastal cities were designed to meet the challenges of the past. Now climate change presents an unprecedented challenge, both present and future, and while cities have started to grapple with it, our urgent work has just begun. New Orleans shows us that we cannot wait until the next disaster to begin planning. We must be proactive and guided by the principles of equity and inclusion. Honolulu reminds us that this work will be an iterative process spanning decades. The decisions we make now will have to be revisited in the future, when new ones will confront us. New York City reminds us to strive for positive transformation. The possibility of a community-driven adaptation project that provides multiple benefits can be achieved.

In order to meet the challenge that climate change presents, cities must factor climate risks into every action and investment starting now. We can no longer think of adaptation projects as something separate; rather, every land use decision, every infrastructure investment, and every housing development must take current and future climate risks into account. We must confront the uncertainty within the range of scientific projections, come to terms with new planning horizons that are much longer than the ones we are used to, and build new governance systems that are equipped to propel interdisciplinary solutions.

One of the biggest challenges will be figuring out how we pay for this work. Too often, cities can access the resources needed to jump-start meaningful adaptation efforts only *after* a major disaster. This means cities are reacting to a problem that requires a deeply proactive approach. We need policy reform at the federal level to provide funding for adaptation and to incentivize action. And we can't afford to wait.

My daughter was born on the day of that late-December nor'easter. How we prepare for climate change will shape whether she is able to live in the same coastal cities that I have. My work, our work, to adapt and build resilience forces a reckoning with the warnings that science is offering. The future of coastal cities is unknown—ours to write. Perhaps we will harness their dynamism, diversity, and grit so our kids, for a while at least, can also call these cities home.

I want to live on a planet that can hold us. I believe we can all still help it, us, do so. If nothing else, why not try? Why not hope, and then act as if? This is our one wild, lone home; what other choice do we have?

—R. O. Kwon

Buildings Designed for Life

AMANDA STURGEON

During the twentieth century, humans became "inside creatures." More than 90 percent of our time is now spent indoors: eating, sleeping, learning, working, socializing. We have largely ignored this growing separation from nature, as well as the significant contribution it's having on climate change. In the United States, 40 percent of energy goes to buildings—to heat, cool, light, cook, and power—and most of that is from fossil fuels. We adopt energy-efficiency measures while overlooking a root cause of energy addiction: separation from nature. Buildings embody our perception of nature as "other," something to destroy or dominate. By rethinking the built environment, we can reclaim a relationship with nature and find remedies for the climate crisis, at the same time. Having a thriving, positive, and hopeful relationship with the world we inhabit could be a path to addressing the climate crisis.

Many buildings are made of too much glass yet without operable windows; they overheat like greenhouses, then guzzle energy to cool down. Large square structures house dark interiors with little natural light, then tap the electrical grid to light them. We ship construction materials across the world, ignoring local alternatives and the amount of carbon emitted. This is the norm for the built environment today, given an almost singular focus on maximizing square footage and return on investment. We can upend this status quo by remembering that buildings are actually, rightfully, human habitats.

My vision is this: When any building is created, it emerges in response to the unique climate, ecology, culture, and community of its specific location. The structure becomes a story of that place, and it embraces the same natural breezes and sunshine to heat, cool, and light the space that have been used to make buildings comfortable for centuries, even millennia. When any existing building is renovated, it reintroduces daylight and natural ventilation and opens up to the outside. Building in this way can significantly reduce energy consumption and emissions, ultimately opening the door to achieving zero-carbon buildings that use no fossil fuels during construction, renovation, or operation.

It's a simple concept—connecting to nature in the built environment—and for thousands of years it was the only way we knew how to build. In every part of the world, people have infused architecture with plant and animal motifs; incorporated gardens, ponds, and atria into buildings; and brought the outside *in* by keeping plants and animals close. Indigenous homes worldwide have long been heated and cooled naturally, using the evaporation of water to cool spaces in desert climates, for example. Traditionally, buildings were an expression of their people and unique to their place.

The movement to reclaim this way of thinking is called biophilic design—*philia* meaning "love," *bio* meaning "life." Rooting the design of the built environment in a "love of life" is about much more than daylight, fresh air, and views; it is a strategy to reawaken hopeful and positive connections between people and nature. Our buildings shape us; they express and inform our values, cultural beliefs, and economic stature. Because the climate crisis calls for all aspects of our society to transform, our buildings will have to look, feel, and function fundamentally differently too.

At a busy restaurant, the only tables free are in the middle: We prefer to have our backs to a wall to view the outdoors. If you watch where people hang out, patterns emerge: in the sunshine, around one another, along the edges of open spaces and shorelines. We can't escape our intuitive craving for comfort, safety, and security. We are instinctively connected and responsive to the living systems that have evolved with us. Yet our buildings sever and ignore this innate connection.

Many other intuitive patterns of behavior that are present when we are immersed in nature can be incorporated into our built environment: the exhilaration, fear, and awe we experience at the edge of a waterfall; the excitement of exploration and discovery we feel when we uncover something new; the peace and tranquility that arise from being in nature, with its variation of light, space, texture, and pattern. Our buildings can evoke all of these emotions and experiences, allowing us to be happier, healthier, and more productive. Biophilic design is starting to transform our workplaces from dingy beige cubicle farms into spaces full of sunlight, fresh air, and color; and starting to revolutionize our hospitals and schools, where the young and the sick too often spend their days in lifeless, dark rooms. It is no surprise that transformations of space precipitate transformations of creativity, healing, and learning too.

Just one hour in nature has been shown to improve our memory and attention by 20 percent. Imagine the feeling of lying in a park and watching the clouds drift by, the awe of discovering newly hatched chicks in a bird's nest, the exhilaration of riding waves at the beach. Those experiences clear our minds, leave us refreshed in spirit, and fill us with a renewed sense of possibility. Imagine that connection to nature is commonplace, everyday, in the midst of growing cities and the urban lives that more and more people are adopting, rather than only on weekends or when we are on vacation.

It may be hard to picture now, but buildings of the future will eliminate heating and cooling systems for most of the year and rely on natural ventilation and daylight. Already a combination of very high efficiency and passive solar can maintain comfortable indoor temperatures with outdoor temperatures as low as 45 degrees Fahrenheit. The buildings of the future will remove the line between inside and outside, and zero carbon will be the norm rather than the exception.

Te Kura Whare is the first cultural center created in generations for the Tūhoe Māori in New Zealand. As I describe in *Creating Biophilic Buildings*, the center's biophilic design reflects a renewed vision for a future not saddled with the colonial devastation of the past. The building is created from local materials: wood harvested from the Tūhoe lands and mud bricks created from the earth as part of a job training program. The inside spaces vary in volume and scale, mimicking the natural patterns of the region's vast forests. Light creates sensory variation as the sun casts dramatic shadows that move across rooms throughout the day and seasons. Natural ventilation and openings to the outdoors allow the spaces to be full of the sound of wind and the laughter of children running in and out of the central café. The building is able to represent and reinvigorate the deep connection that the Tūhoe people have with their land, with one another, and with all living species.

Resources and programs are on the rise to support a broad transition toward biophilic building. The U.S. Green Building Council's LEED rating system now references biophilic design, and it is included in the WELL Building Standard and as one of the imperatives of the Living Building Challenge. Building owners, designers, and engineers are deepening their knowledge, and increasingly, universities are incorporating biophilic design into curricula so that the next generation of professionals get the training they need to shape the buildings of the future.

Biophilic design is the underlying design solution for addressing the impact of the building sector on climate change. It is also an essential opportunity to change our relationship to one another and to all of life on Earth. Implicit in the choice we make about the built environment is a choice about ourselves: Are people separate from nature, or are we part of nature? I believe we are part of nature and that we are intricately connected to all other living species. It is essential that we start creating habitats for humans to thrive in, instead of developing buildings purely to make a profit.

If we design new buildings and redesign existing buildings to celebrate that human-nature connection, we will have rooms filled with dappled light and shadows. We will feel the breeze on our skin and hear the sounds of birds and the rustling of trees when we are inside. In essence, our indoor spaces will make us feel as vibrant and healthy as we do when immersed in nature outdoors. We will develop a deeper connection to ourselves as earthly beings while inside, while lightening the impact that our human-made habitats have on the world.

The Straits

JOAN NAVIYUK KANE

Ledum, Labrador Tea, *saayumik*.
A matted growth beneath the most shallow
depth of snow on record in all our winters.

Pausing upbluff from the edge of ice
I broke from branches leaves to pin
between my teeth and tongue

until warmed enough for their fragrant
oil to cleanse you from me.

Somewhere in a bank of fog
beyond the visible end of open water,
alleged hills were windfeathered—

drainages venous. In routes
along the shore forever slipping
under, I am reminded—in the city

one finds it simple to conceive nothing
but a system, and nothing but a world of men.

Catalytic Capital

Régine Clément

> The plain truth is that capitalism needs to evolve if humanity is going to survive.
>
> —Rose Marcario

In August 2019, leaders of thirty-three certified B Corporations, legally bound to consider social and environmental impacts in their business decisions, put their names on a full-page ad in *The New York Times:* "Let's get to work . . . to make real change happen." Rose Marcario of Patagonia, Kat Taylor of Beneficial State Bank, clothing designer Eileen Fisher, and others directed their message to the Business Roundtable, an association of some 180 CEOs of large U.S. companies. The Roundtable had just made an announcement that making money for shareholders should no longer be the sole objective of a corporation—the B Corp ad was an invitation to them to walk the talk and join efforts well under way.

The confluence of crises we face prompts deeper questioning about what got us here in the first place and how we might get out. Put simply, we are a society overly driven by capital and wealth; many of our values and mindsets are rooted in the desire for money. Social identity is in many ways calculated by wealth and channeled through our consumer purchases and employment. At the same time, rapid global economic growth has led to mass inequality. And it's all entangled with climate change. The time has come to rethink the relationship among our economy, social progress, and ecological systems.

To address what brought us here, to a place where our climate is in crisis, I find it necessary to think in a mindset of *systemic transformation:* departing the unsustainable practices of extraction, production, and consumption that have supported blind economic growth, while moving to a set of practices that recognize the long-term value of a sustainable economic system, which will allow us to maintain a livable climate.

The scale of the challenge we face today should not come as a complete surprise. Philosophers and socioeconomists have reflected on humans' desire to accumulate wealth and referenced our destructive trajectory time and time again. In her 1913 book *The Accumulation of Capital*, Polish revolutionary Rosa Luxemburg critiques capitalism as an inherently destructive system, engaged as it is in constant, imperialist expansion to acquire raw resources and access ever more labor and new markets. According to Luxemburg, economic expansion and the resulting devastation of the environment and communities is not a defect of capitalism but an inherent feature of it. Climate change is the most extreme manifestation of that and requires us to address it not only urgently but also systemically.

The challenge of climate change is perhaps best defined as our challenge to end destructive capitalism.

Within that challenge sits another puzzle: How can we use the mechanics of capitalism as it currently exists to transform it? Luxemburg would almost certainly be skeptical of this possibility, but advancing climate solutions requires money: We need to unlock and reallocate capital on the order of trillions of dollars a year. According to the Intergovernmental Panel on Climate Change, transforming our global energy system will require annual investment of roughly $2.4 trillion through 2035, and additional capital will be needed for other critical sectors, such as agriculture, industry, and transportation.

One effective place to start is private capital, specifically the capital of ultra-high-net-worth families. Single-family offices—the private investment firms of wealthy individuals and families, typically with at least $100 million of assets under management—manage an estimated $5.9 trillion of assets globally. Their wealth alone clearly won't be sufficient, but the collective influence of investments by these families has the potential to inflect capital markets toward decarbonizing our economy.

In 2016 I became determined to find a way to activate this pool of investors. I left my career as a public servant to run CREO, a network of U.S.-based family offices aiming to accelerate capital into climate solutions. Motivated by the belief that the transition to a low-carbon economy has to be commercially viable, these families and their investment managers are among the trailblazers of the surge in climate investing. Four years ago, CREO was a network of about twenty-five family offices with about $50 billion in assets under management; today we work with more than one hundred family offices and other aligned investors, who manage roughly $500 billion and invest over $1 billion annually in climate and sustainability solutions.

The mechanics of CREO are straightforward: We *bring investors together* to share investment theses and insights; we *fill in knowledge gaps* by conducting research on new investment strategies; we *build relationships* among aligned investors to accelerate their collaboration; and we *provide investment opportunities* that enable the community to act on what they learn. By equipping investors new to the field with insights from experienced climate investors, while making sure experienced investors are kept informed of the fast-evolving climate-investing marketplace, we are growing the pool of competent professionals. By shaping investment behavior and catalyzing capital within the CREO community, we aim to shift behavior and capital more broadly, using the power of peer-to-peer influence, network effect, and new investing norms.

I get that this might seem contradictory. On the one hand, I write about ending destructive capitalism to address climate change. On the other, I suggest that climate investing by wealthy families, all of whom have benefited in some way from the destructive nature of the capitalist system, is a critical lever to address this existential problem. I also suggest this knowing that the world's richest 1 percent have carbon footprints that are 175 times higher than the poorest 10 percent. Economic and social inequality, as well as what has been dubbed "carbon inequality," must all be addressed in our work to stop climate change and, by extension, to end destructive capitalism. At the same time, we can use the reality of our current system to transform it: highly concentrated wealth, investors seeking returns, the ability of the wealthy to exert influence, and investment successes attracting much needed capital to climate solutions.

One of the key advantages of climate investing by family offices is that they are nimble; they can more readily take and manage risks related to new technologies or unproven business or financial models. For example, without wealthy families' early and high-risk investments into First Solar, today a publicly traded solar panel manufacturer, the United States might not have any significant presence in that industry. Early family-office investments into QuantumScape, a pioneer in solid-state battery technology, are helping Volkswagen transition to making electric vehicles. Beyond Meat, Impossible Foods, and Ripple Foods all benefited from investments by family offices, precipitating the growing shift to plant-based meat and dairy alternatives. Beyond direct investments, family offices are often early investors in funds, such as those created by Generation Investment Management, and create their own funds, such as the Vision Ridge Partners Sustainable Asset Fund, expanding the talent pool in this field and proving the financial models. Through their risk tolerance and early investments, families have already paved the way for larger institutional investors to invest billions of additional dollars into climate-focused companies and funds.

To be sure, we cannot limit climate solutions to single technologies, companies, funds, or business models to "correct" our unsustainable economy. For example, to invest in, or incentivize investment in, renewable energy does not automatically decrease emissions; that clean electricity generation could simply end up layered on top of fossil fuels. Investing in renewables should be paired with other systemic measures such as implementing a carbon price through federal legislation or imposing a moratorium on coal. That's why some wealthy families are also putting their money toward campaigns for political and policy change.

If we are to address the climate crisis and destructive capitalism, we must question the underlying rules and tools that direct the behavior of capitalism. The failure of the market economy to price so-called externalities, such as the positive impacts of natural ecosystems and the negative impacts of greenhouse gases, should lead us to question the limitations of how we measure financial performance: an individual company's EBITDA (its earnings before interest, taxes, depreciation, and amortization are taken into account) and GDP (gross domestic product) to assess broader economic growth, for example. Both metrics lack a holistic perspective, failing to consider externalities and thus rewarding and reinforcing destructive capitalism despite its harms to people and the planet.

So we need big, systemic change, but in my experience with CREO, transformation can also start small. It begins with shared *intention*, a well-orchestrated *community* approach, and creating the necessary *competence*. Every day I see elements of transformation growing through collaborative work; the creation of new solutions, new strategies, and new investments; and a growing number of investors committed to building their expertise so they can move quickly from intention to action.

Building on the momentum of climate investing to date, we must go from billions of dollars invested in climate solutions to trillions this next decade—because that's what systemic transformation will require. We have seen only the tip of the iceberg of conventional investors, such as BlackRock, waking up and joining the climate marketplace. But despite the meaningful work and expertise developed by trailblazers, the financial industry as a whole is ill prepared to apply a robust climate lens to financial decisions and has fast-growing needs for good data, investing frameworks, and competency building.

Movement of capital at the scale required cannot be done without conventional investors' participation and real change in the financial industry. Wealthy families and their family offices should continue to pioneer, to leverage their assets, strengths, and influence. Our still-emergent field of climate investing should also support all asset managers to transition thoughtfully, quickly, and successfully. We need a larger, stronger team to move forward.

Let's adopt a transformation mindset and keep Luxemburg's critique in mind. In addition to catalyzing capital, let's find the courage to address the underlying impediments to decarbonizing the economy and to develop a system that considers socio-ecological-economic progress over the long term—not merely economic growth for now and for the benefit of the few. We must bring our moral compass to capitalism and bring its destructive elements to an end.

We need business goals that are bigger than doing less harm or emitting less carbon. The goal has to be what the planet needs: a climate fit for life. And we need optimism and courage to get there—maybe more than we need science or data.

—Erin Meezan

Mending the Landscape

Kate Orff

My hometown, Crofton, Maryland, is typical of many middle-class developments of the 1960s, with its wide streets, single-family houses, two-car garages, and, of course, a golf course. A frontier suburb on former farmland, it was a pleasant, considered, and rather unremarkable setting. It was also a local harbinger of global change: car-centric logic increasing auto emissions, standalone homes demanding large loads of electricity, and sprawl accelerating habitat fragmentation. During my lifetime, our neighborhood pond stopped freezing over for ice skating in the winter, became choked with algae in the summer, and eventually dried up for two years straight, filling with bird and fish carcasses and alarming local residents.

Fueled by subsidized gasoline and publicly funded road networks, the carbon-intensive backdrop of "suburban sprawl" has reached its zenith in the United States and has been replicated around the world for the last few decades. While planners and landscape architects were busy laying out suburbia, the world's ecosystems were being degraded, polluted, and torn apart.

Habitat fragmentation and toxic pollutants created the biodiversity crisis; sprawling development, which discouraged community life, contributed to a social crisis; and exploding greenhouse gas emissions precipitated the climate crisis.

These three interrelated crises will define the planning and design professions for the next century. To be a landscape architect today is to wake up each morning with sorrow and guilt, as well as a sense of mission. My goal as a design teacher and practicing landscape architect is to reimagine my profession as a form of collective gardening.

At the same time that Frederick Law Olmsted and Calvert Vaux—two prominent early adopters of the term "landscape architecture"—made their iconic mark with Central Park, the salt marshes of Jamaica Bay were being used as New York City's trash heap. Highways and bridges came much later; shoreline gradients were concretized; islands were impounded into straight landmasses for two airports; and sewage flowed freely for decades. The recent revitalization of Jamaica Bay shows how to mend and love landscapes moving forward.

Jamaica Bay was once thick with cordgrass and teeming with shellfish and flounder. But the shellfish population collapsed during the twentieth century from overharvesting and raw sewage, and in recent decades marshlands that once cushioned against extreme weather started to slough off into open water, succumbing to pollution and rising sea levels. By the 1990s, the landscape was drowning. Local activists, birders, and conservationists banded together to track and publicize the marshes' imminent disappearance and habitat loss for hundreds of species.

Federal, state, and local action converged (at times, prompted by lawsuits) in response to this landscape emergency. New York City and New York State acted to address pollution and improve water quality. The U.S. Army Corps of Engineers and the National Park Service embarked on an island-building process that reused dredge spoils to replenish starved marshlands, spraying on layers of sediment. Local volunteers began planting cordgrass, critical to salt-marsh ecosystems, and "oyster gardening" to reseed the shellfish and mussels and stabilize the sand that, together, hold cordgrass roots in place.

I've learned four lessons from Jamaica Bay, now a beloved living landscape, that can inform our collective gardening and landscape architects' climate action: visualize the invisible; foster ecosystems as infrastructure; create participatory processes; and scale it up.

Visualize the Invisible

The active agents of the destruction of natural systems are often largely *invisible.* We can't see the excess of nitrogen in our bodies of water that concentrates to create vast hypoxic dead zones, nor the clouds of carbon dioxide pumped into the air from our tailpipes, power plants, buildings, and industries. Imperceptible pesticides used to create that coveted bright-green lawn are killing native pollinators and contaminating the soil beneath our kids' feet. These are all critical factors precipitating the sixth mass extinction and threatening our food supply. But more than all of that opaque ruin: We can't love what we don't see.

Petrochemical America, the book and traveling exhibit I created with Richard Misrach, was an effort to break through ecological invisibility and foster empathy. We visualized the lower Mississippi, a place in the crosshairs of oil extraction, environmental injustice, and sea level rise, through a series of maps, photographs, and drawings, alongside profiles of actions for change and riverbed restoration. These visualizations armed local advocates, including the Louisiana Environmental Action Network, the Louisiana Bucket Brigade, and RISE St. James, for discussions with chemical companies, state government, and the media; they have also been displayed in museums around the United States to spark dialogue on energy and climate change.

This is just one example of the way illustrations can help us make imaginative leaps, as well as leaps of consciousness. Visualization can usher the heart and mind into the insidious impacts of chemicals and fossil fuels, the widespread loss of intertidal ecosystems, the disappearance of interior forest species, the steelhead trout's struggle to spawn upstream, or salinized rice fields no longer fit for growing. It can also usher us into contemplation of alternative futures.

Foster Ecosystems as Infrastructure

Thriving landscapes are next-generation climate infrastructure: generous riverbanks, healthy reefs and mangroves, protective dunes, and living shorelines. Traditional "beautified" landscapes bear little resemblance to the biodiverse, food-producing, coast-protecting, and carbon-absorptive ecosystems we need, and too often a reductive mindset keeps us focused on technical solutions, overlooking natural ones. Conventional design and engineering can actually greenwash climate risks. We normalize and render "safe" the dynamic corridors of rivers. We fortify waterfronts and literally pave the way for developments atop used-to-be wetlands. These practices normalize the occupation of land not suited for habitation. Wetlands in Harris County, Texas, barrier dunes in Mobile Bay, Alabama, and riverbanks in Ellicott City, Maryland—these are all places where water has historically been but where we don't want it to go now. So we mark them "flood-prone" and move on, even as storms grow more intense and waters rise. Instead, landscape architects and planners need to focus our full force on designing, building, and engaging people around protective ecosystems.

After Superstorm Sandy devastated the Northeast, my design firm, SCAPE, participated in a yearlong process called Rebuild by Design. With oysters and juvenile striped bass as our imaginary clients, our team developed a concept called Living Breakwaters, rocky protective structures, seeded with oysters, to reduce wave action along the eroding shore of Staten Island. These structures rebuild the three-dimensional mosaic of coastal habitat that protects us from extreme weather, while the affiliated Billion Oyster Project and its learn-by-doing science curriculum bring education to the shore. This "oystertecture" project is equal parts habitat reconstruction, climate risk reduction, and community engagement.

Whereas Olmsted and Vaux brought nature into the heart of the city, it is now critical to rebuild the environmental context broadly, for the entirety of human habitation at a global scale. Design and careful maintenance are needed to reconstruct coastal reefs, diversify our farm fields, and reforest our uplands. Now is the moment to reframe them as necessary regenerative infrastructure that helps clean our air, filter our water, protect our coasts, provide our food, and stabilize our climate. Thickened forests along waterways to reduce erosion, leaf litter and decay in our yards for native pollinators, tangled mangroves to secure our shores, solar panels and wind turbines mingling with meadows—these will all be part of a reset, climate-responsive landscape. Mending our ecological infrastructure and pairing it with renewable energy should become the highest priority for the profession of landscape architecture.

Create a Participatory Process

Communities are often suspicious of change, and local laws sometimes prohibit even the most modest solar panel or geothermal well. Yet we know the climate crisis demands massive transitions in energy, land, industry, transport, buildings, and cities, which have direct impacts on the places where we live, work, and play—the places we love.

The public needs to be informed and empowered to consider the trade-offs and transitions ahead, and communities must be engaged in codesigning the biodiverse, low-carbon, climate-conscious physical world we want to manifest. Distributed energy systems and public transportation need to integrate into the fabric of our cities. Rising seas will displace entire communities and urban heat may make some zones uninhabitable. Therefore, a primary task is to design participatory, educative, and fun processes for community engagement. Our democratic trust is wearing thin—we need mechanisms for developing a generous dialogue among landscape architects and planners and communities, so change happens in a way that is equitable and just.

SCAPE is advancing a democratic, participatory process in Metro Atlanta for the Chattahoochee RiverLands project—a hundred-mile vision plan to reconnect the region to its river, its source of water, recreation, and wildness. That plan includes a physical trail that parallels the Chattahoochee River, bringing access to riverbanks and options for nonmotorized transport. We have held almost one hundred stakeholder and community meetings and educational events, including after-school creek crawls and weekend river paddles, in addition to formal stakeholder convenings, all aimed at engaging Atlantans in reimagining their relationship to the river. Rather than conceiving a "master" plan at arm's length, we are workshopping and co-creating with residents, building on decades of local leadership for the river.

This new mode of working—a more public role of organizing communities and choreographing ecological repair—is needed to correct the idea of nature as something that exists outside of human agency. The Earth is both a physical setting and a decision-making commons that must be cultivated. Communities that invest in, engage with, and, yes, love their landscapes are at the heart of making our way in a climate-changed world.

Scale It Up

From individual oyster to functional reef to healthy bay lands to vast blue ocean—the idea of scale is embedded in living landscapes. There are multiple scales of human action too, from the garden at your doorstep to a restored forest at the edge of town to regional watershed policy to national legislation and beyond. We need all of them to mend the Earth.

It's time to redesign the mighty Mississippi as a living river system, reconnecting the river with its floodplain and coastal basins. The Dutch have implemented the Room for the River program along the Rhine, aimed at giving the river more space and ability to cope with high volumes of water. A future Mississippi River National Park could take a similar approach, to reverse the U.S. Army Corps of Engineers' practice of concretization and control and its unintended consequences. Iowa pig farmers, Arkansas bird-watchers, and Louisiana oystermen would all benefit from letting the river reconnect with its shoals in the upper watershed and from nourishing the Mississippi Delta with land-building sediment. An aspirational project like this could knit the center of our country together and reclaim the river as a recreational, linear park rather than a waste canal. The Mississippi helped build this country, and we need her help again in the climate crisis.

Our nation's coastlines also call for bold aspiration. Today the National Flood Insurance Program (NFIP) provides funds to rebuild flooded properties exactly as and *where* they were, often putting people back in harm's way. Rather than being an instrument for equitable adaptation, disaster response, and managed retreat from rising rivers and seas, it exacerbates the income divide by giving payouts to already-wealthy homeowners, sometimes for vacation homes. What's more, the NFIP is already over $20 billion in debt. In its place, we need a federal buyout policy that enables families to choose proactively to move. Research has shown that many people want to and feel proud to leave their land in buyout programs and are willing to give it back to Mother Nature. Let's embrace a fresh vision for an interconnected and publicly owned American National Shoreway, which could be made possible by encouraging retreat, funding equitable relocation, and rebuilding protective shorelines as linear parks that maximize public access at the water's edge.

Mending the fabric of the physical landscape at a local scale, as in Jamaica Bay, or at a regional scale, as a Mississippi River National Park would, shows us a way forward. Reviving our nation's green-blue infrastructure is the collective work of our time. Stitching ecological connections is good for vibrant and healthy communities, animals with migratory lifelines, food systems supported by living soils, and the global carbon cycle and climate. Mending some things will also require tearing up and dismantling others: equitably unbuilding places in harm's way, depaving roadways, blowing up dams, ripping out concretized stream channels, jackhammering asphalt roadways, and cutting down sterile seawalls that push risk downstream. Acts of design can tend to the rights of the channelized river, the clam in the muck, the climate migrant to survive and thrive.

It's time to get our hands in the mud. Let's actively love and mend our messy, swampy, dusty, busted-up landscapes—the tide pools for darting crabs, dark forests for scarlet tanagers, dead trees for owls and bats, thick grassy dunes for nesting plovers. Tending to and dwelling among our living landscapes can start small: plastic pickups, piling up logs for habitat, gardening with oysters, and pulling invasive ivy from that patch of tulip trees down the block. We face a global landscape emergency. Let's knit what we can back together.

5. PERSIST

MADELEINE JUBILEE SAITO

We Are Sunrise

Varshini Prakash

When I first learned about the climate crisis as a teenager, I would lie awake late into the night, imagining what it would mean for me and people who looked like me around the world. I couldn't get the images out of my head: what people would do to each other when faced with no food, no water; the cages and guns that would greet people seeking sanctuary in other countries. I felt alone, small, powerless.

So many young people around the country and the world have grown up seeing our political system failing us. For twice as long as I've been alive on this planet we have known about the climate crisis. And for just as long, the wealthy and powerful have knowingly profited off pollution, lied about the science, choked our democracy with their dollars, and stolen our futures. Our generation is living at the crossroads of life or death.

My family is from the southern tip of India—a state called Tamil Nadu. My dad grew up on a huge stretch of white beaches. He'd tell me stories about eating fresh mangoes under palm trees; the delicious food my grandma cooked; playing cricket in the sand with a dozen other skinny little Indian boys. In December 2015, a deluge of storms hit the state's capital city, Chennai, causing one of the worst floods the region had ever experienced. Roads were engulfed; children and mothers walked miles in chest-deep water to find sanctuary. The air reeked of rotting flesh and the water sat stagnant for weeks, brewing disease. Thousands were displaced and hundreds killed. My *patti* called to tell me of the damage to their apartment and the suffering the flood had caused. By sheer luck, my grandparents were away during the storm; otherwise, they too might have been stranded with no food or water or electricity for days. My nightmares are still thick with stories like this.

Growing up as a short, Brown girl, I felt that the whole world of politics and elections and the government itself reveled in my exclusion. But I fell in love with social movements in college. I was taking classes in environmental science when a friend asked me to emcee a demonstration against new fossil fuel infrastructure in western Massachusetts. I agreed, but I was super nervous and felt nauseous for days. But when my friend finally handed me a megaphone and I walked out to a crowd of a hundred students, I almost burst into tears. For the first time in my life, I felt like I wasn't just a small person facing the climate crisis alone. I was powerful and I had people with me. And then I really got active.

In the dead of winter, I attended a demonstration in Washington, DC, with forty thousand people to protest the construction of the Keystone XL pipeline from Canada's oil fields through the American heartland. I helped kick-start a divestment campaign at my university to pull all endowment dollars out of fossil fuel investments. I quickly fell in love with organizing because it's about changing politics, not just lightbulbs, and challenging the status quo—and because building a social movement means practicing democracy beyond the realm of elections and political parties, in ways that are vibrant and inclusive.

During the Obama years, young leaders and activists like me witnessed the emergence of a national movement to stop climate change and saw the beginnings of federal climate policy. At the same time, we read one scientific report after another warning that the crisis was more severe and urgent than we could have imagined. We needed a movement in America that could bring together millions of young people to build power and make climate action, rooted in racial and economic justice, a priority in our nation.

Out of this awakening, Sunrise Movement was born. A dozen climate activists from across the movement landscape, fed up with the status quo and the existing green organizations, decided to build something far more ambitious, justice based, and youth centered. We spent over a year doing strategic planning before we launched in spring of 2017. We developed a three-part theory of change to build power, spark massive government action, and drive wholesale transformation. Here's what guides our work.

First, People Power—an Active and Vocal Base of Public Support

Political scientists Erica Chenoweth and Maria Stephan conducted a groundbreaking study of resistance movements of the twentieth and early twenty-first centuries. One key finding is that it takes just 3.5 percent of a population getting active—voting, donating, taking to the streets, talking to their neighbors—for a campaign to win. At the current U.S. population, 3.5 percent would be about 11.5 million people—roughly the number of people who live in Ohio.

Already more than seven in ten Americans understand that climate change is happening, more than six in ten are either "alarmed" or "concerned" about it, and a majority want government and corporations to do more to address the problem. We have a ton of passive support for this issue. Now we need to translate more of it into *active* support: people voting on climate, donating to campaigns, active on social media, signing pledges, calling their elected representatives, feeding and housing organizers, and supporting in other creative ways to reach that critical 3.5 percent.

Second, Political Power—a Critical Mass of Deeply Committed Public Officials

There is a deep failure of political leadership in our country, and our leaders' cowardice, at a time when courage is needed most, directly impacts our well-being. The 2016 election was the breaking point: Trump won the presidency as the House and Senate tilted to a climate-denying GOP. Fossil fuel interests took hold within the administration: Rex Tillerson, the former CEO of ExxonMobil, was appointed secretary of state; Secretary of the Interior Ryan Zinke started opening up federal lands to oil and gas extraction; Andrew Wheeler, a coal lobbyist, was put at the helm of the Environmental Protection Agency.

We realized that people power without political power does not suffice. We need allies in office and must help them get there. Otherwise we're just railing against a group of elected officials who aren't accountable to our values or our communities. We have to make sure that our politicians win or lose based on where they stand on this issue.

Third, the People's Alignment—Social, Economic, and Political Forces United Around a Shared Agenda

The United States has seen two major political alignments in the last century. The first was the New Deal alignment of President Franklin D. Roosevelt, which started in the 1930s and lasted into the 1970s. It was defined by an active government passing sweeping social policies that helped support and elevate working Americans. The second was the alignment of President Ronald Reagan, initiated in the 1980s, which focused on government as the problem and the market as the solution. It brought a new set of values around individualism, disinvestment in the public sector, and deregulation of industry that persists today.

As we face entangled crises of inequality and climate change, we need a new political alignment in America—a people's alignment—that espouses policies and programs derived from values of equality, fairness, and safety for all people, regardless of the color of their skin or how much money they have or where they live. We need to build the people's alignment alongside progressively minded think tanks, social movements of all types, government officials, faith institutions, and more to usher in a new era of policy making and transformation in this country.

In November 2018, two hundred Sunrise activists walked into Representative Nancy Pelosi's office to deliver photographs of our loved ones, tucked inside envelopes stamped with the words "What is your plan?" Our demand: for the Democratic Party to back the Green New Deal (GND) and swear off campaign money from Big Oil. As we filled the hallways and Pelosi's office and images of the Camp Fire raged on the office TV, my friend Claire Tacherra-Morrison recounted how fires had consumed her aunt and uncle's California home. Representative Alexandria Ocasio-Cortez, whom we had worked to elect, joined us in solidarity.

In the days that followed, thousands of articles were written about climate change and countless youth took action; that action helped transform climate change from a political afterthought into a top issue in our nation's politics. Three months later, Representative

Ocasio-Cortez and Senator Ed Markey released a landmark GND resolution that raised the bar for federal climate action above anything we'd seen before. As the Democratic presidential campaign kicked up, candidates made GND endorsements and eschewed fossil fuel money. And Sunrise exploded from just twenty local chapters in November 2018 to a national movement with more than three hundred chapters across the country. We're big, and we need to get bigger.

We are also seeing momentum at state and local levels. In 2018 Sunrise endorsed Chloe Maxmin's bid for the Maine House of Representatives. She ran on clean air, water, and energy and turned her rural district blue. Representative Maxmin quickly pulled together labor and environmental partners to pass a Green New Deal for Maine that boldly commits the state to 80 percent renewable energy by 2040. In New York in 2019, Sunrise and other grassroots organizations helped Democrats win back the state legislature and secure passage of one of the most ambitious climate policies in the country.

We're working to make the necessary climate action politically possible, and then to turn the politically possible into the politically inevitable. Sure, we're going to have to compromise; there will be disappointments. But we believe we'll get further if we hold our ground.

Sunrise is fundamentally changing the way we talk about solutions to the climate crisis. For so long, climate action has been framed as taking something away from people. The GND is about the opposite—providing people with millions of good jobs, reinvigorating our economy, and putting money back in the hands of working people. It's about alleviating inequality between different groups of people. It's about ensuring we have clean air and clean water. *And* it's about stopping climate change.

Too often, people talk about the climate crisis like it's something middle- and working-class people need to bear the financial brunt of, when one hundred fossil fuel producers are the source of more than 70 percent of emissions since 1988. So when we're talking to people whose wages have stagnated for the last forty years while ever more wealth goes to people at the very top, they're right in asking, "Why should I pay for this?" We cannot address this problem without talking about labor, jobs, health, and equity—they are inseparable from climate and are top concerns for our people.

A July 2019 poll showed that overall 63 percent of Americans support a GND. When you break that down, there was support from 86 percent of Democrats, 64 percent of independents, 55 percent of rural voters, and 40 percent of White evangelical Christians. For young people (ages eighteen through thirty-eight), it's 77 percent. The GND could actually be a winning vision for America.

Within Sunrise, we use song to build joy, necessary joy. We use it in times of fear or intensity as a way to show solidarity with one another and show our strength. We use it in times of sorrow or pain to give voice to our feelings. We use it in moments of anger. Like many movements throughout history, we use song to bring people together and give voice to what we're here to do.

The stakes are so high. As a movement, we are intimate with the implications of our own failure. They sit very heavy on us. If Sunrise navigates this political moment poorly, millions of people could die. But if Sunrise navigates this moment well, millions of people will be elevated out of poverty, and we can help protect human civilization as we know it.

There always will be the fear of defeat, but there is also the knowledge that something is more important—a deep spiritual calling toward doing something to better people's lives, the lives of those we love and those we'll never meet. That's our shared purpose in the face of uncertainty. The only failure would be to do nothing at all.

I look to texts for wisdom and guidance in this moment as well. One of my favorites is the Tao Te Ching. There's one line that I keep returning to: "Do your work, then step back. / The only path to serenity." The way I see it, if we are constantly striving, constantly putting every ounce of ourselves into making this world better, that is a life worth living.

Choose to fight only righteous fights, because then when things get tough—and they will—you will know that there is only one option ahead of you: Nevertheless, you must persist.

—Sen. Elizabeth Warren

At the Intersections

Jacqui Patterson

One of my earliest memories is visiting my dad's homeland—Jamaica, West Indies. I remember dancing to reggae on the beach with my brother, embracing the sun and sand. From that moment I have loved Jamaica's wind, water, and rhythm; its unique nature, culture, and heritage; its ties to sub-Saharan Africa, "the Motherland."

I also remember the White tourists surrounding my brother and me, taking pictures as we danced. They assumed that we were native Jamaicans and presumed that in buying their plane tickets, they had bought the right to own images of Jamaicans to later display for their enjoyment and other people's entertainment. I didn't know then, basking in the innocence of childhood, that I was having one of my first experiences with the subtleties of racism.

Exposure to racism, in subtle, overt, and systemic ways, has been a consistent thread in my life.

Fast-forward to college, when, through reggae, I began to explore themes of revolution. With Bob Marley in the ears of my Walkman, I was constantly encouraged: "Get up, stand up!" I was reminded: "None but ourselves can free our minds." I was invited: "Won't you help to sing? These songs of freedom."

Upon graduating with a degree in special education, I joined the Peace Corps, which brought me back to Jamaica. Living there from 1991 to 1994, I saw the vestiges of colonialism still very much in play. Whether it was to the tourism industry or the wealthy White people who lived on the island, Black Jamaicans were pervasively in servitude. It was the epitome of *Black Labor, White Wealth* as penned by Dr. Claud Anderson.

I lived in a community outside Kingston called Harbour View, which was polluted by a cement plant, and the surrounding area had an array of health problems due to exposure to particulate pollution.

The water supply in Harbour View was contaminated by industrial waste. The community pointed to the neighboring Shell Oil refinery as the culprit, but the only compensation they received was a few ventilated pit latrines and a handful of dollars for a recycling program at the elementary school.

At this point in my life, I was all too familiar with injustice—from White kids chasing me back across the tracks with their taunts of "N***er, go home!! You don't belong here!!"; to watching my brother, as a young Black male in Chicago, joining the ranks of endangered species and being challenged by gang forces to become the hunted or the hunter; to volunteering at a homeless shelter at the start of the AIDS epidemic; to working in shelters for domestic violence survivors. I had gained a deep sense of racism, poverty, misogyny, and their intersections.

But this experience in Harbour View was the first time I saw all of that intersect with the environment. I saw firsthand how environmental justice became an issue as corporatocracy began to dominate Jamaica through the bauxite industry, the cement industry, the banana industry, and beyond.

Out of travesty the Community Environmental Resource Center was born. With this project, in addition to addressing a plethora of environmental justice issues, we began to discuss climate change impacts, which had particular resonance as the people were still recovering from Hurricane Gilbert, which had slammed the island in 1988.

Hurricane Gilbert gave Jamaica a taste of what is to come with the increase in frequency and severity of extreme weather linked to climate change.

Jamaica has social vulnerabilities heaped on top of environmental vulnerabilities. Wealth inequality, historically high rates of unemployment and debt, and the second-highest murder rate in the world are added to natural hazards such as hurricanes, earthquakes, floods, droughts, and fires, creating chronic crises for the island. According to one analysis, Jamaica is the Caribbean island most vulnerable to future climate change and is also highly vulnerable economically.

Trying to analyze and understand the geopolitical and socioeconomic situation there, I started a book club. We read books such as

How Europe Underdeveloped Africa, *The Spook Who Sat by the Door*, and *Invisible Man*, which explore themes around international development, globalism, and racism. Underneath the "irie" facade, characterized by the "no problem, mon" catchphrase, festered the ongoing exploitation of this beautiful nation by those seeking to capitalize on its bounty of natural resources, including cultural richness, with zero regard for those who lived there.

I decided to study public health and community organizing because I wanted to be a part of changing the systems that resulted in such deep inequities and substantial vulnerability.

I pursued a master's degree in social work and wrote my thesis on the role of race in the high incidence of infant mortality in African American women. I examined the role of neighborhood indicators on birth outcomes in Baltimore, a predominantly African American city, and helped carry out the Baltimore Fetal-Infant Mortality Review. We documented what is still true today: Where you live is a powerful indicator of health and well-being, due to social, economic, and environmental factors, such as air pollution. Place matters.

Out of grad school, in 2001, I returned to international work with a faith-based organization called IMA World Health, working primarily in sub-Saharan Africa. If I thought Jamaica suffered from post-colonization, it paled in comparison with what was happening in the Motherland. Extraction of natural and human resources was so extensive that many places were practically uninhabitable. Industrialized, wealthy nations cornered the market on imperialism, domination, and exploitation, wreaking such havoc that countries were torn asunder by poverty, war, pollution, and health epidemics at scales equivalent to genocide. Meanwhile, the World Bank and the International Monetary Fund triggered such oppressive austerity measures that already economically hungry nations were being starved.

I was in the belly of the beast of a failed development paradigm.

In the Democratic Republic of the Congo, resource wars raged, with women used as weapons, while industrialized nations established trade and manufacturing policies that guaranteed that factions in developing nations would continue fighting over scraps. In Malawi, where I worked on HIV and AIDS, people were doubled up in hospitals, one person sleeping on the floor below another person sleeping on a bed. In Zimbabwe they had to dig graves twelve feet deep so that there was space to bury another person six feet above, because they had run out of grave space for those dying of AIDS. Meanwhile I was part of a program that insisted on buying only antiretroviral drugs that were manufactured in America while there were drugs being produced by India that were literally 1,000 percent cheaper. And in South Africa pharmaceutical corporations were suing the government for legislating to reduce drug prices. They were raking in profit while Black people died by the hundreds of thousands.

I quickly came to understand why healthcare activists referred to many in the pharmaceutical industry as "genocidally evil."

Five years later, during a goodbye party on my last day at IMA World Health, my eye was drawn to a TV screen: Billowing from a window was a sheet with "SOS" written on it. As I watched the wrath of Hurricane Katrina assail the communities of New Orleans, I again saw the impact of government neglect and failing infrastructure, with African American lives once again being most impacted. I went to the Gulf region for six weeks, experiencing firsthand the intersectional realities of climate change and specifically what it means for marginalized communities.

Working at a disaster recovery center (DRC) in Houston, I heard and saw so much: a man with AIDS who didn't have access to his antiretroviral drugs and feared that he would become drug resistant, rendering his treatment regimen ineffective; a woman with a hearing impairment who relied on her young son and me to attempt to interpret for her so that her complex housing, health, and childcare needs could be fulfilled; a man searching desperately for his wife and children; a woman who had been sexually assaulted in the disaster's aftermath.

I was one of the only people of color working at the DRC, and I keenly saw the biases against those seeking refuge. One day, as I came for my shift, I approached the door and was flanked by help seekers eager for the doors to open. The person staffing the door thrust out his hands in a repelling gesture and roared, "GET BACK!" as if commanding a herd of rampaging animals. Another day, a DRC worker, who was a White returned Peace Corps volunteer, announced that, based on his experience working in Nigeria, half the folks coming to the center were likely to commit fraud. At one of the shelters, I encountered a gender-nonconforming person who was forced to identify as male or female for housing and use of the restroom, thus leaving them at risk for abuse or worse. And then we heard of the Black people, seeking to go back into New Orleans to search for food and possibly stranded relatives, being shot on the Danziger Bridge.

What I saw post-Katrina continues to haunt me. That experience also motivates me to support community resilience ahead of disasters and to advance systemic disaster equity during and after emergency situations.

I experienced the impacts of biases against those seeking refuge with only the clothes on their backs.

My next move was an intense stint doing international work, this time as a human rights activist focused on gender justice and its intersections with finance, violence, HIV/AIDS, and climate change. In a focus group of women in Mpumalanga, South Africa, the participants expressed their need for female condoms. As a result of climate change drying up waterways, girls have to walk farther to get the water needed daily for their households. The likelihood of sexual assault is so extreme, it is better if the girls wear the condoms whenever they go to fetch water. A woman who left her native land of Cameroon because the crops had dried up in her community was raped at a border crossing and became HIV positive as a result. These stories drew my tears.

There is a pandemic of devastating impacts at the intersection between violence against women and climate change.

Issues of injustice are wrapped up in each other, inextricably. In frontline communities, places that feel environmental harms first and worst, people are choking on pollution; mothers are burying their babies; children are losing their fathers to incarceration or state-sponsored violence; people are being driven from their lands; hurricanes are decimating tropical islands like Barbuda, making them nearly uninhabitable; sacred ground is being desecrated; oceans are being suffocated by plastic; koalas are burning in fires; and so much more. All while fossil fuel companies and others continue a reckless pursuit of profit and power, and while politicians act at the behest of those seeking wealth rather than those just trying to survive. It is all too clear that justice is not possible in a capitalist system predicated on there being winners and losers, a system rooted in racism, sexism, and xenophobia. This is the system that has put us on the path to catastrophic climate change.

Now, as the senior director of the NAACP's Environmental and Climate Justice Program, I am often asked why the NAACP would be working on environmental issues and climate change. People express bemusement that a civil rights organization, founded in 1909 out of the struggle for equal rights of Black people in the United States, would have a program focused on climate. My response is that with our mandate to advocate on the behalf of the oft voiceless, the reasons are innumerable. The most polluting coal plants are disproportionately located in communities where residents are predominantly low income or people of color. Burning fossil fuels is resulting in sea level rise, which is displacing communities from Kivalina Island, Alaska, to Isle de Jean Charles, Louisiana. Shifting rainfall patterns and extreme weather are disrupting agriculture, deepening existing food insecurity in too many communities. Climate change is absolutely a civil rights issue.

In the words of Dr. Robert Bullard and Dr. Beverly Wright, pioneers of environmental justice, we have the "wrong complexion for protection."

One of the first things I did upon joining the NAACP was to develop a document entitled "Intersections" to illustrate how climate change is a threat multiplier for all of the racial-justice issues we work to address and how all of these issues are inextricably connected. We can't have racial justice when our homes don't have the infrastructure to deal with increased heat; when we don't have the wealth to withstand the damage and displacement of climate disasters, and disaster capitalism is rampant; when, instead of being aided in postdisaster contexts, our communities are often criminalized; when prison labor is used to fight increasingly frequent and dangerous wildfires.

We must make fundamental shifts: from a society that drills and burns to power our communities to one that harnesses the sun and the wind; from a society that buries or burns our waste to one that recovers, reuses, and recycles; from a society that genetically modifies, trucks, and ships food to one that advances local production of food that is nutritious and accessible for all. We must have a radical transformation from extracting, polluting, and dominating policies and practices that negatively impact our communities to regenerative, cooperative systems that uplift all rights for all people, providing sovereignty, wealth and asset building, and local control for our communities while nurturing and preserving the environment upon which we all rely.

Before, I was merely a theoretical revolutionary. Now I recognize deep in my heart and soul that the only path to liberation for Black folks and all oppressed people is through revolution—total systems change.

In the course of my work with the NAACP, my awakened self has returned to the places of my childhood on the South Side of Chicago, where personal life and work come full circle and merge into the epitome of intersectionality. I discovered that I grew up within ten miles of three coal-fired power plants, all of which received an F grade in the report *Coal Blooded.* This report, coauthored by the NAACP, Indigenous Environmental Network, and Little Village Environmental Justice Organization, graded plants according to their proximity to communities of color and low-income communities. Coal-fired power plants spew carbon dioxide, sulfur dioxide, and nitrogen oxide—all greenhouse gases—as well as mercury, arsenic, lead, and particulates. This pollution is tied to everything from respiratory illnesses and birth defects to learning disabilities, attention problems, and violence.

Is this why so many of the kids in my school had asthma? Have asthma-related absences, combined with attention problems, put kids in Chicago on the path to the school-to-prison pipeline? Is continuing to burn cheap, dirty coal in pursuit of profit responsible for the record-breaking heat wave of 1995, which killed 739 of my neighbors?

My work today centers on supporting frontline communities to advance their visions of liberation; it focuses on building power to actualize systematic change. Recognizing that the same forces driving systemic oppression also cause climate change, our program has a comprehensive agenda to address the roots of inequities while equipping communities to weather the storms (literally and figuratively) in the meantime. Among other things, this requires policy change, narrative shift, and leadership development. We want to ensure that opportunities in the new energy economy are equitably shared and build healthier, wealthier, and more democratically governed communities. Instead of a school-to-prison pipeline, we need a Black-green pipeline.

What inspires me is the fact that the revolution has already begun.

Across the country and world, community by community, people are building microcosms of the systems and societies we need to reverse the tide of catastrophic climate change and become a world that respects all rights for all people, in harmony with Mother Earth. Communities in Pittsburgh are growing their own food. Indigenous people in North Dakota are resisting pipelines and holding on to their lands. In Tallahassee, Baltimore, and Detroit, communities have united to defeat incinerators, both existing and proposed. Youth are suing the federal government for ruining the climate they will have to live in. Women are standing up for gender justice, getting elected into office, and growing cooperative enterprises from goods production to local food and beyond. People are campaigning to get money out of politics and into communities. Across the nation, people are taking ownership of their energy systems. Groups like the Pan African Climate Justice Alliance, La Via Campesina, and the Indigenous Environmental Network are speaking truth to power at the United Nations' annual meetings on climate change. In short, resistance is rising and systems are changing.

How do we scale the transformation that's already under way? First, communities must uphold a vision of the desired future; that is a necessary precursor to lasting change. Then, political education leads to effective strategizing. Frontline communities and populations know the solutions to environmental and climate injustice and must be decision makers. Demonstrating the benefits of policies will strengthen commitment, passion, and the effectiveness of advocacy. We must embrace localism and organize to build power to stop the bad and build the new. And we must shift the narrative to foster the context and broad political will for change.

That is our theory of change. That is how we will overcome our rapid progression toward further catastrophic climate change and all of the other converging crises that assail communities nationwide. We must link arms across movements, across race, across class, across organizations.

We channel Sister Assata Shakur in chanting:
"It is our duty to fight for our freedom.
It is our duty to win.
We must love each other and support each other.
We have nothing to lose but our chains."

Every year I spend a month in Jamaica over the holidays. On the last visit, in the winter of 2019, I took long walks by the water every day. I ate healthy food straight out of the ground, listened to reggae music, and got work done. But as I gazed out at that gorgeous cerulean sea, I began to despair—that unless we turn around the runaway inequality train and get society on track, Jamaica's days are likely numbered in the face of catastrophic climate change. It made me weep with anticipatory grief.

So I came home, home to the movement for environmental and climate justice with all its beauty and fire and hope. Jamaica is worth fighting for, as is every nation, community, family, and child.

Together we can. Together we must.

Did It Ever Occur to You That Maybe You're Falling in Love?

Ailish Hopper

We buried the problem.
We planted a tree over the problem.
We regretted our actions toward the problem.
We declined to comment on the problem.
We carved a memorial to the problem, dedicated it. Forgot our handkerchief.
We removed all "unnatural" ingredients, handcrafted a locally-grown tincture for the problem. But nobody bought it.
We freshly-laundered, bleached, deodorized the problem.
We built a wall around the problem, tagged it with pictures of children, birds in trees.
We renamed the problem, and denounced those who used the old name.
We wrote a law for the problem, but it died in committee.
We drove the problem out with loud noises from homemade instruments.
We marched, leafleted, sang hymns, linked arms with the problem, got dragged to jail, got spat on by the problem and let out.
We elected an official who Finally Gets the problem.
We raised an army to corral and question the problem. They went door to door but could never ID.
We made www.problem.com so You Can Find Out About the problem, and www.problem.org so You Can Help.
We created 1-800-Problem, so you could Report On the problem, and 1-900-Problem so you could Be the Only Daddy That Really Turns That problem On.
We drove the wheels offa that problem.
We rocked the shit out of that problem.
We amplified the problem, turned it on up, and blew it out.
We drank to forget the problem.

We inhaled the problem, exhaled the problem, crushed its ember under our shoe.
We put a title on the problem, took out all the articles, conjunctions, and verbs. Called it "Exprmntl Prblm."
We shot the problem, and put it out of its misery.
We swallowed daily pills for the problem, followed a problem fast, drank problem tea.
We read daily problem horoscopes. Had our problem palms read by a seer.
We prayed.
Burned problem incense.
Formed a problem task force. Got a problem degree. Got on the problem tenure track. Got a problem retirement plan.
We gutted and renovated the problem. We joined the Neighborhood Problem Development Corp.
We listened and communicated with the problem, only to find out that it had gone for the day.
We mutually empowered the problem.
We kissed and stroked the problem, we fucked the problem all night. Woke up to an empty bed.
We watched carefully for the problem, but our flashlight died.
We had dreams of the problem. In which we could no longer recognize ourselves.
We reformed. We transformed. Turned over a new leaf. Turned a corner, found ourselves near a scent that somehow reminded us of the problem,
In ways we could never
Put into words. That
Little I-can't-explain-it
That makes it hard to think. That
Rings like a siren inside.

Dear Fossil Fuel Executives

CAMERON RUSSELL

Dear ExxonMobil, Shell, BP, and Chevron executives,
Dear chairmen of the boards,
Dear Jamie Dimon and the board of directors at JPMorgan Chase,
Dear fossil fuel corporate leadership and investors everywhere,

When I was sixteen, I found oil in my backyard, metaphorically speaking. That was when I started modeling and, before I knew it, found myself in the infamous Victoria's Secret show, modeling for Prada, and on the covers of *Vogue* and *Elle*. I felt ethically dubious cashing out. Modeling rewarded something I hadn't worked for, my looks, and required that I participate in reinforcing all the superficial ways we value and devalue human beings. For all my misgivings, I didn't stop. Seventeen years later, modeling is still my primary source of income.

Because of this, I feel like I know you. Or maybe I am you. This much is true: My teenage metaphor was more literal than I had imagined. It turns out we both work in highly extractive industries that pay a small number of us extremely well at the expense of people and the planet. And because of our similarities, I am writing to you with the compassion and respect I have for myself—something I didn't always have. In fact, I spent years examining why I made certain decisions and learning how to hold myself in dignity as I found a path forward.

Today, like you, I am aware of the damage wrought by my industry, and yet I still work within it. Fashion's carbon footprint is big and * growing, responsible for 8 to 10 percent of global emissions. We must accept responsibility for growing a culture of rampant consumerism, too: Three-fifths of all clothes end up in a landfill or incinerator * within a few years of being produced. By some estimates, textile production is on track to use at least a quarter of the world's carbon budget by 2050.

I was on set in 2013 when reports came in that a Bangladeshi garment factory in Rana Plaza had collapsed, killing at least 1,134 people and injuring more than 2,500 others. That factory produced clothing for brands including Benetton, the Children's Place, Joe Fresh, Mango, Primark, and Walmart—some of which I had worked for. I was on a shoot three years later when I read *The Guardian*'s reporting on the thirty-eight locals charged with murder in connection with the collapse. I absorbed the unfolding tragedy that somehow my coworkers and I, intimately connected through profit, had escaped any responsibility for—not one person from the brands whose clothes were made at Rana Plaza was indicted. Although there were efforts to improve safety in the years that followed, progress was feeble. The five-year commitments made by major North American and European retailers to improve worker safety expired. When *The New York Times* asked Mahmudul Hassan Hridoy, who runs a support group for Rana Plaza survivors like himself, what this would mean for workers, he said, "All will return to how it was when Rana Plaza happened."

Six years on, a *Wall Street Journal* investigation found that Amazon, which had surpassed Walmart as the biggest clothing retailer in the United States, was selling garments from dozens of the blacklisted, unsafe Bangladeshi factories. In 2019, when fifty thousand Bangladeshi garment workers went on strike demanding higher wages, the fashion press didn't give the workers so much as a blog post of coverage.

Maybe you can't relate to what I do for a living, but I believe you may understand the horror I felt that day in 2013. Maybe it was a similar feeling; maybe your breathing became shallow, too, during the 1989 *Exxon Valdez* oil spill, or at the public's horror when it became clear that Exxon management had not only known about the climate repercussions of fossil fuels since the 1970s but had gone to great lengths to cover them up and, further, to wage a massive misinformation campaign to instill doubt about global warming. Maybe you were around in the 1980s when Shell was widely boycotted for being the principal supplier of crude oil to apartheid South Africa. Perhaps you still cannot sleep when you recall the apocalyptic 1991 burning of the Kuwaiti oil fields. Maybe you stood in front of a television in 2005 and saw people standing on roofs after Hurricane Katrina. Possibly you were shaken when the Gulf Coast was hit again, this time by the 2010 BP Deepwater Horizon oil spill. Or when the strongest typhoon ever measured—2013's Typhoon Haiyan—left nearly ten thousand people dead in the Philippines.

Now most things that jerk us awake to climate chaos are mainstream media's daily headlines. We know there are mere years left to avoid a climate change that most humans will be unable to adapt to. When you let these truths in, do you wonder quietly, desperately, *What can I do?* Despite our different life paths, perhaps we have this in common.

Like me, perhaps at some point you were, and maybe still are, a hard worker, a force to be reckoned with. Perhaps you, too, went to an Ivy League school, graduated with honors. Our privilege has set us up to be powerful voices in nearly every room we walk into. You have probably experienced that intoxicating, galloping high of being commanding, clever, and clear, an effective leader, a smooth operator. Like me, you have likely found it easy to ignore all the things you don't (or do) know in these moments, and some of this willful ignorance has in fact allowed us to sprint to the front, where there is another intoxicating reward: money.

Motivated by your own ambition, or by your children, or by your faith, I believe that you, too, have at least thought about how to do better. But as the years wear on, we have also witnessed a growing number of tragedies that force us to confront our continued complicity in industries failing humanity.

As leaders in our industries, we can cast about for a new commitment, a change we can make—something disciplined, material, concrete. A new hire, a new policy, create a sustainability office, allocate, donate, but *what else*? We tell ourselves that if we didn't have a seat at the table, we couldn't take even these small steps. Comforted by this thought, we can believe that what we have done is more than enough—foolish even—in the face of the impossible.

Still, the situation continues to worsen, the clock ticks, the carbon count rises. Are you scrambling to understand what enables the extractive nature of our industries? Are you curious?

In my industry, some of the most public efforts to make fashion sustainable address only natural-resource impacts—for example, working to use recycled or organic fibers. While they're a necessary piece of the puzzle, these solutions have generally ignored the multitude of systemic factors that make fashion's extractive business model so possible and so profitable. We find a deeper analysis in the work of Céline Semaan, who traced today's fashion trade routes alongside colonial shipping routes and discovered they were mostly the same. Or in the work of academic Minh-ha Pham and reporter Anne Elizabeth Moore, who observe that the vast majority of women who power the fashion industry, as 80 percent of its workforce, are women of color and most don't make a living wage.

So we must ask: How are sexism and racism, colonialism and imperialism, helping make the highly extractive parts of the fashion industry flourish? Without careful attention to systemic causes, fashion sustainability means nothing more than finding ways to sustain business models, and has little to do with sustaining people or the planet.

Again, these challenges must feel familiar to you. The fossil fuel industry is also enabled by many broken systems. In the United States alone, direct and indirect subsidies exceed annual spending by the Pentagon, propping up the profitability of oil, gas, and coal. It is enormously difficult to incorporate long-term thinking when short-termism dominates boardroom strategy, as the market encourages quarterly growth above all else.

Then there's the industry's long history of using environmental racism to avoid detection and regulation. A study by the NAACP found that Black Americans are 75 percent more likely than Whites to live in "fence-line" communities next to commercial facilities whose noise, odor, traffic, and emissions directly affect the population. In real numbers, this translates to more than 1 million Black Americans living within a half mile of oil and gas wells, processors, and compressors, and 6.7 million living in counties with oil refineries. This proximity and exposure increase the risk of cancer, lung conditions, and other illnesses. And that's just the United States. The world over, your industry relies on disenfranchisement to avoid the democratic policy making that should have long ago forced an end to such injustices.

The more I learned about my industry, the worse I felt. In fact, I stopped working for a while. Have you ever done that? It is a privileged response to despair. I think we often allow ourselves to be ignorant. Not ignorant of the issues—you know where this all leads, and maybe you ogle the bunkers of billionaires or have built your own—but ignorant of another way forward.

Once I began to accept the overwhelming truth of our systemic failures, I saw clearly that the invisible majority of fashion workers at the economic, social, and political margins are actually the people who make fashion culture and usher in change, who have been creating alternatives for decades, even centuries. In the 1830s, for example, * the Lowell Mill girls, young garment workers, organized the first women's labor union in the United States. In the wake of the 1911 Triangle Shirtwaist fire, organizing by garment workers and their allies led to the New Deal labor protections Americans still enjoy today.

Today, thredUP, the world's largest online thrift store, is building a logistics backbone to carry upward of $50 billion in retail value. Taking a more decentralized approach, the UK start-up Depop is growing a platform that allows sellers—mostly stylish teenagers—to sell their own curated secondhand clothes. Depop reports over fifteen million users. Fibershed, in California, farms regeneratively to produce fiber for garments. After two decades in the business, designer Tracy Reese shut down her original brand and relaunched an intentionally small, Detroit-based sustainable line. "I want to ship fewer collections," she told *Vogue*, "because the world just does not need so much merchandise." In New York City, Custom Collaborative trains and supports women from low-income and immigrant communities to launch fashion careers and businesses. Ninety percent of their products are made from repurposed, upcycled textiles.

The list goes on. Thrilling alternatives exist everywhere.

The same is true in the energy industry. As soon as you stop centering fossil fuels, stop centering yourself, you may see a way forward. Technological advancements mean that in 2020, solar and onshore * wind costs will consistently fall below fossil fuels. As in fashion, incredible alternatives are being developed in the economic and political margins, everywhere. Native Renewables, for example, is working to bring affordable off-grid solar energy to fifteen thousand Navajo and Hopi homes in Arizona. (Native American and Alaskan Native * tribes have the highest rates of homes without electricity in the United States.) In New York, PUSH Buffalo is helping low-income communities access solar power and weatherize their homes and is developing energy-efficient affordable housing. If some of the least-resourced communities are finding ways to build a just and sustainable future, then certainly fossil fuel companies, with their vast power and resources, can find ways to transition with fairness to the people they employ. Now is the time for us to throw our outsized weight behind brilliant alternatives, many emerging from smaller players and independent voices.

Since the Paris Agreement, banks around the world (JPMorgan Chase is the worst offender) have invested $1.9 trillion in fossil fuels, growing oil and gas companies. The money and power you all have amassed should *immediately* be poured into dismantling extractive systems and into clean, renewable energy. Simply put, there can no longer be companies that extract or profit off fossil fuels. Transition looks like shifting investments, retiring infrastructure, and retraining workforces.

The fact that our industries are comparable shines a light on a sad reality—most big industries rely on some level of extraction to continuously grow their profits. In fashion we must also retire business models built on excess consumption. The production of garments doubled from 2000 to 2014, and the average consumer bought 60 percent more—but kept each piece for half as long. We must develop a just supply chain that produces not just garments but culture and community as sustainable outputs.

But I realize the comparison can only be taken so far. While fashion faces the difficult task of dramatically retooling and cutting production, your industry is even more challenging. Energy requires even more fundamental change. So, yes, I'm rather brashly suggesting you take on something harder and, frankly, more necessary.

When we take a wider view—beyond fossil fuels, beyond consumption culture—we can see ways for all of us to walk dignified, in community, into the future.

I now see my participation as a lever. If I stay in fashion, I can reallocate resources, focus, and time; I can meaningfully opt out, and I can organize. For the past few years, I have helped build Model Mafia, an active network of over four hundred model activists around the world. We have found shared values through the realization that this job, often focused on superficial success, offers us the opportunity to do good for our families and communities. We're committed to radically reforming our industry.

I hope you also find your humanity, your true value. We are alive at this moment. What we do now will help decide the fate of our species and most living things on Earth.

Sincerely,
CAMERON RUSSELL

#MeToo said Mother Earth

—Alex Lieberman (age twelve), climate strike sign

Sacred Resistance

TARA HOUSKA
ZHAABOWEKWE, COUCHICHING FIRST NATION

Sitting on the well-worn wood floor of a tiny cabin just off the Red Lake reservation in northern Minnesota, I breathed deeply. The aromas of drying sage, burning cedar, and bubbling wild-rice soup filled my lungs.

"Would you like to visit with them?" the elder asked. For the past twenty minutes, he had quietly spoken of four medicine drums that made their way back to the people, liberated from behind museum glass, where they sat silenced, isolated for many decades. I nodded and greeted the new spirits sitting with us after they were carefully unwrapped.

On the rough wooden table beside us, several handmade snow-shoes rested, waiting for final touches and feet to hold until the crusty snows of *Onaabani-giizis* ("hard crust on snow moon") disappear.

There is no "climate change" or "economic policy" in this place; its walls hear murmurings of the forest, of the water, of the life around us and our place within it. Interaction with one another, with all that lives, is thoughtful, respectful. Simple discourse is pregnant with hidden wisdom awaiting further reflection. It is in spaces like this that my feet grow heavier, the roots grow stronger.

To step from that warm, clear room into spaces led by westernized values and institutionalized disconnection is jarring, saddening, and maddening all at once. The corporate boardroom, with its lifeless statistics and figures; the environmental coalition, with its budgets and measured campaign outcomes. Here, the voice of the land is sanitized; our Mother becomes a summation of "natural resources" for us to consume or conserve, always at arm's length.

Much of the space we call "the climate movement" appears to be modeled after the same systems of inequity and separation we are attempting to change, undo, or outright dismantle. Massive NGOs stick to traditional advocacy methods, techniques that aren't too disruptive or unfamiliar to the philanthropy they need to ensure hundreds or even thousands of paychecks and programs. Language like "front lines," "grassroots," "youth leadership," and "inclusivity" float over wineglasses at lavish funder gatherings. In such settings, the horizontal patterns of community that build connections and translate wisdom flicker dimly, outranked by the need to know who holds the purse strings and whose name you should know. Connections become transactional or cliquish, hierarchy is entrenched, scarcity mentality is omnipresent, and personal growth is interwoven with westernized "success" as a marker of authority.

So many of us come together to mobilize, make waves, and share space—to feel some agency within a machine that has cogs in nearly every aspect of our day-to-day lives. Everyone looks to one another hoping for the right answer (and also hoping to make that answer first), aiming to deduce the best tactic, the elusive "silver bullet" to stop or at least slow down the crisis that fills us with anxiety. Underlying much of our movement is a fundamental survival mechanism that operates on "me" instead of "we."

Far too much of our collective energy is directed toward a pursuit that leaves us mirroring capitalism, individualism, and that which we fight. Bringing in more people (and ultimately more dollars) seems to be the only acceptable theory of change. Money—the currency of individualism—hangs like a heavy cloud over campaigns calling for systemic change. It is undeniable we are all stuck in the clutches of capitalism and the fossil fuel economy, from the tiny collective to the global environmental NGO. Creating educational materials, conducting research, bringing lawsuits, etc.—it costs money. But there are millions of people who are waiting for direction on how to protect our shared home. These times are indeed urgent, as all of our messaging states. We can do far better to direct the energy of those already engaged into substantive change.

"Okay, load up!" It's the kind of icy cold that snaps and bites the insides of noses. The sun isn't up yet; at this time of the year, daylight is a rabbit dashing through snowdrifts, barely glimpsed and quickly gone. Steamy breath drifts through the air over equally steamy coffee and eggs made by a yawning kitchen crew. Someone bundled head to toe is doing a head count. Massive red pines creak overhead.

An hour later, a young woman is suspended twenty feet in the air, blocking the entrance to Enbridge's U.S. tar sands pipeline terminal. Enbridge is the largest fossil fuel infrastructure corporation in North America; one of its latest efforts is a massive pipeline that would carry more than 700,000 barrels of bitumen sludge per day through my people's territory, through our wild rice beds, on its way to the shore of Lake Superior. A group of us circles the base of the wooden tripod the young woman sits at the apex of, praying for the water and land, chanting our message of keeping fossil fuels in the ground while snow swirls around our bodies. Work has stopped at the facility; the sound of approaching sirens grows over the sound of boots stamping the ground to keep frozen toes from complaining.

Out here in the places few see, the risks are higher and the connections are tighter. There are far fewer questions of intent when you are standing together on a lonely road, wondering if there will be weapons pointed at you over your "Stop Line 3" sign, knowing that at least one and possibly all of you might be jailed—or worse. But we also know the risks we take directly disrupt the status quo and will reach the investors and shareholders who continue to financially back fossil fuel expansion in the face of climate crisis.

Our steady resistance forms cracks in the world of profit margins. It transitions us away from self-destruction. We are a thorn in the side of a world that believes it must extract to exist, a bone-deep reminder there are other ways of being and people willing to take personal risk for something greater than any one individual. I have seen fear in eyes shielded behind riot gear, fear of a braid of sweetgrass and a prayer, of the person who stands unarmed to protect the land.

Some of us have been out here living on the land for years, for a lifetime, for generations, defining our existence by the seasons. The fish and wild rice harvest of summer turns into the harvesting of deer in the fall, into trapping and spearing during winter snows. When the winter breaks and the trees run with sap, we share a fire and boil down maple syrup together. We put our thoughts and prayers into action together—we will defend the land with our bodies, our freedom, our hearts.

Local men arrive at our blockade, shouting obscenities and laughing. One points at a "Missing & Murdered Indigenous Women" sign and tells the woman holding it she might go missing if she keeps showing up out here. The last time land defenders were at this location, two men driving by passed a gun back and forth in plain view. Corporations spend millions on campaigns to sow distrust and hatred toward anyone who disrupts the status quo. Native people have been targets since the beginning of colonization. Uneasy racism and fear of the unknown "savage mind" are constants.

After hours of back-and-forth, the police decide to start sawing the legs of the tripod, the most dangerous way to bring the young woman down. A newly arrived ambulance innocuously waits down the road.

When the saw's teeth begin cutting through the pine poles, the tripod wobbles ominously. She cries out as the crowd below rushes forward, shouting at law enforcement to stop. I see red as I try to speak clearly and calmly in an attempt to reach these people who would directly, intentionally harm a human being for the profit margin of an oil company.

In moments like these, I think of the safety that numbers can bring. I wonder if police would be so emboldened with thousands watching instead of a dozen. I wonder what would happen if the environmental movement truly stood with the land it speaks of, side by side with impacted communities who bear the brunt of the climate crisis. I wonder about the dedicated industry effort to criminalize direct action of concerned citizens, concerned human beings. I wonder about the militarized and established financial influence over law enforcement to surveil Indigenous peoples teaching others to live off the land and reclaim the values we hold within. We must be onto something—if not by the fact that Indigenous peoples hold 80 percent of the world's remaining biodiversity, then by the sheer scale of the opposition we face.

I wonder if most people truly believe we will achieve change comfortably.

At the end of this day, our friend lowered herself out of fear of tumbling to the pavement. As she was led to the squad car, the Indigenous people of the territory thanked her for standing up for the land, for the water, for us. Above her windburned cheeks there were tears in her eyes. "It was an honor to stand with you," she said, after we scraped together enough money to bail her out of jail.

Those windswept prayers and fierce eyes sit in my heart when I am in rooms of so-called power. The rooms writing laws and policies, the rooms of slick language and tacit jousting of tense allies with the same financial backers. In spaces like these, I become a "frontline representative," an uncomfortable, pitiable reality alongside conceptual data presented in more familiar language, with more familiar solutions.

The smell of sap, boiling over a fire while a story is shared in a copse of maples, sits just past my fingers as I slide a lawsuit across the table to a corporate banker. Some of us leave the land to bring our case to the financiers of the industry we oppose, to present the data and oppositional testimony the banks ostensibly have no knowledge of. These rooms pump filtered, sharp air across Earth's elements manipulated into steel and glass. In here, I feel like an exotic bird to be examined for potential danger. In here, alongside discussion of financial investments, I remind corporate heads that they drink water and breathe air. As awkward as it can be to remind a person of their own humanity, it has proven exceedingly effective to bring Indigenous rights and the voice of the land into these spaces.

Banking institutions across the world are divesting from fossil fuels in a growing chorus of change. Land defense poses deep reputational risk; human rights abuses and lack of tribal consent precipitating lawsuits can mean years of project delays and mounting costs. Indigenous advocacy brings a collision of values across a polished desk.

In westernized society, we are indoctrinated with self, with the pursuit of comfort, from the moment we are born to the moment we die. Birth costs money. Life costs money. Death costs money. The advent of social media has simultaneously connected us to one another and disconnected us from one another. Listening takes a backseat to whoever is speaking the loudest. Climate advocates, those brave souls who stare down mass extinction, walk a fine line: succumbing to ego and maintaining a comfortable living versus ever-evolving efficacy and self-assessment. Ours is a world tinged with doubt.

Quite simply, I do not believe we will solar panel or vote our way out of this crisis without also radically reframing our connection with our Mother.

As the movement latches onto whatever its latest source of momentum is, I suggest we do far, far more to address our values—the core principles that guide us. Currently, we demonstrate the "value" of our work by the outcomes achieved. While many of us share a desire to protect our shared homelands, we often fail to consider what *values* we are endorsing along the way.

Across many Indigenous teachings and cultures, several threads weave in and out, creating similar patterns and core understandings. Each person has their lifetime to learn these threads, these values, to discern the truth through ever-more-complex ways of knowing. These are values centered in balance, in life, not in human fallacy. They are values that require commitment and self-sacrifice. Humility, recognition of fragility and our place in the circular framework of nature, empathy, courage, respect—those are just a few threads of the many values that are meant to guide us through the whole of our lives.

Many Indigenous cultures have ceremonial practices of physical deprivation or intense bodily challenges. To be humbled by the lived knowledge that our bodies cannot survive without water is to move water from the conceptual into the actual. To suffer through weakness in the footsteps of our ancestors is to grasp at our infinitesimal place in the world, to feel those who came before and the weight of those yet to come. These practices are not a one-time lesson but a way of life. We must relearn these truths over and over again. Human beings fall into the pitiful because we are just that—beings with fragile egos and self-doubt. As visceral as lived knowledge can be, the fast-paced nature of modern society can quickly dull it.

Taking a stand with the land is far more accessible than many realize, and I hope far more of us do it—with our bodies, our choices, with whatever we can give to truly espouse and teach values that reflect life. The simplicity of traditional foods from a local ecosystem can feed a soul; the decision to build community and live a life of empathy, balance, and humility is a values shift badly needed.

The sacred is all around us, always. The sacred is in each of our bodies, a miracle of life and water and earth. It is present in every object we touch, every wall and window we somehow believe separates us from our Mother. We each hold the beauty of creation in every fiber of our being. We are never far from the answer to the problem we have created—it is within each of us.

On the Fifth Day

JANE HIRSHFIELD

On the fifth day
the scientists who studied the rivers
were forbidden to speak
or to study the rivers.

The scientists who studied the air
were told not to speak of the air,
and the ones who worked for the farmers
were silenced,
and the ones who worked for the bees.

Someone, from deep in the Badlands,
began posting facts.

The facts were told not to speak
and were taken away.
The facts, surprised to be taken, were silent.

Now it was only the rivers
that spoke of the rivers,
and only the wind that spoke of its bees,

while the unpausing factual buds of the fruit trees
continued to move toward their fruit.

The silence spoke loudly of silence,
and the rivers kept speaking,
of rivers, of boulders and air.

In gravity, earless and tongueless,
the untested rivers kept speaking.

Bus drivers, shelf stockers,
code writers, machinists, accountants,
lab techs, cellists kept speaking.

They spoke, the fifth day,
of silence.

Public Service for Public Health

GINA McCARTHY

All my life, I felt government service was the place I could best make my mark on the world. I spent thirty-five years working in local, state, and federal government. It was a system that made sense to me, as frustrating as it could be at times: We used science to identify and mitigate risks to human health and natural resources through a well-defined public process. I was proud to design programs, implement regulations, and enforce laws that are fundamental to protecting people's health and well-being. And in all those years, I never once questioned whether any role other than public service could be more productive, fulfilling, or important.

Then came the Trump administration, with its unique brand of "public servants," many of whom seemed to relish the prospect of dismantling our democracy from within. I ran the U.S. Environmental Protection Agency during President Barack Obama's second term and have seen the current administration try to tear apart the very work I helped spearhead. There are one hundred regulations in limbo or rolled back, and counting, including the Clean Power Plan to reduce greenhouse gas emissions from power plants; the Clean Water Rule to protect rivers and streams that one in three Americans relies on for drinking water; and rules to reduce exposure to toxic chemicals in our air and in our homes that lead to premature deaths and illnesses. This administration has outright threatened government employees—especially our scientists. And to top it off, it denies the reality of the climate crisis that science tells us is not only real but gaining in urgency as we struggle to change the way we fuel our economies worldwide.

Imagine watching a president devote himself to unraveling not only your work and your legacy but your efforts to make the world a better place for your kids, for everyone's kids. It has been gut-wrenching. As I watched, I grew increasingly convinced that the future of our democracy was in question, along with our rule of law and our children's ability to live in a healthy, safe, and sustainable world.

And like so many people who were shaken into action by the Trump administration's assault on essential protections for vulnerable people, I decided to act. I committed to "leave it all on the field" in order to stop the unprecedented rollbacks of environmental protections.

To the surprise of family and friends, I made the shift from a career in government to take the helm of a leading environmental advocacy organization, the Natural Resources Defense Council. At NRDC we fight regulatory rollbacks and help put in place cost-effective solutions at the state and local levels to reduce greenhouse gas emissions and improve public health. We also know that's not enough. Only a movement can motivate the breadth of action we need to confront climate change.

I remember when my first child was handed to me after I gave birth. I looked at my son Daniel and simply could not believe that at first glance I fell more in love with him than anyone I had ever met. And in some ways it terrified me. From that moment forward, my future and my happiness were no longer about me; they were dependent on the health and well-being of my son, and it was my responsibility to protect him as best I could.

I am also fortunate to have two daughters, Maggie and Julie, and two grandbabies to love and cherish. The youngest members of our family will be only thirty-two and thirty-one years old in 2050—the year when science tells us we must be well on our way to a zero-carbon future if we hope to avoid the most destructive impacts of climate change. Before my grandchildren were born, 2050 seemed far away. Not anymore. They are the faces of climate change for me. It's that personal.

It's also professional. After decades of experience, I cannot envision our government finding a pathway to address climate change without a broad and demanding push from the people. That's just how democracies work: It's how our democracy worked when we tackled America's pollution problems in the 1960s and 1970s. Earth Day happened, the EPA was created, and bedrock environmental laws were passed. We reduced air pollution in the United States by 70 percent while GDP tripled.

Responding to climate change requires a range of actions at all levels—not just a price on carbon emissions but laws, regulations, and the right targeted incentives; technological advances; divestment from fossil fuels and reinvestment in solutions; coups by enlightened shareholders; winning litigation against fossil fuel companies; and winning elections to put climate champions in power. It requires shifting budgets from dirty energy to clean and shifting agricultural practices to improve soil health. It requires phasing out fossil fuel–based fertilizers, pesticides, plastics, and other synthetic chemicals that have imposed immeasurable damage on our health, our ecosystems, and the biodiversity that keeps our planet alive. Each one of these efforts is necessary, and success will not happen without people demanding and embracing them.

We cannot be fooled into thinking there is a shortcut by which our system of democracy can easily fight climate change—that we'll act just because we have seen the science and know the threat posed to our health, environment, economy, and national security. Democracy is a government "of, by, and for the people." It is not a dictatorship, and it is not a spectator sport. Without broad-based demand from local communities, states, businesses, and nonprofits, our government cannot succeed. We know this; our ancestors knew this.

So it's time for folks like me, who look at the public process as the cornerstone of our democracy, to step up. It is time to stop focusing on what government *can* do and start recognizing the critical role we all play in *making* government do its job.

Here's my plan: I'm going to communicate the stakes of this crisis and the opportunities that await us if we get it right. I'm going to make sure that everyone understands the connection between climate change and health. I'm going to lift up the women who are already doing this work and whose ideas and energy will be essential to building the world I want for my grandchildren. I'm not going to let myself get discouraged or burned out.

I'm not just talking about the threats of pollution; I'm talking about building the future we want to see: a future that is clean, healthy, more just, more sustainable—and no longer reliant on fossil fuels. Let's tell the complete climate story: the scale of the crisis and the promise of the solutions. Yes, most people know that climate change is real and it's scary. They just haven't tried to do anything about it.

Maybe it's too big. Maybe they don't know what to do or how. Maybe they think it's just about "the environment," collapsing glaciers, or polar bears rather than our families and our children's future.

Health provides a compelling reason to act and a clear way to measure and celebrate success. It personalizes climate change and opens a window to talk about all of the ways climate actions will make our lives better and help us move toward a more thriving and equitable world.

Exposure to extreme temperatures can cause heatstroke and death. Altered weather patterns cause crop failures that lead to malnutrition. Burning fossil fuels fills the air with sooty particulates, causing lung and heart problems. Warmer and wetter temperatures lead to the spread of mosquito-borne diseases like Zika, West Nile, and dengue, and tick-borne diseases like Lyme. Sixteen percent of all premature * deaths across the world are the result of exposure to air pollution—almost nine million human beings annually, more than those killed by tuberculosis, malaria, and AIDS combined. It's serious business. It matters and it should motivate all of us to act.

And climate change is not an equal-opportunity killer. It goes after the most vulnerable among us: children. Heat stress and air pollution can lead to preterm labor and increased risk of low birth weight. Zika can pass from mother to fetus, causing deadly birth defects. Children inhale more pollution in proportion to their body weight, which can have permanent effects on their development. If you're a parent and your child uses an inhaler, you get it. Four million kids worldwide * develop asthma each year simply because they have the misfortune of living near a major roadway. Shame on all of us who could be doing something about it.

Simply put: Climate change is the most significant public health challenge in the world today. We can reframe climate solutions as opportunities to invest in public health, which will make our world healthier and more just today while we forge a future we can be proud to hand to our children. We'll see immediate benefits like fewer asthma and allergy attacks, safer and more resilient communities, more green spaces for outdoor activities, and fewer traumatic experiences that lead to mental and physical health problems. We'll support communities that have been left behind, communities that are most vulnerable to the impacts of climate change. And if history is any guide, women will be the ones who make it a reality.

Courageous women have been at the forefront of the environmental movement. In publishing her game-changing book, *Silent Spring*, in 1962, marine biologist Rachel Carson sounded the alarm about the dangers of pesticide overuse and transformed our perspective on the natural world. She was not just an early participant in the movement; she pioneered it. She, and others like her, blazed a trail for countless women over the years—scientists, researchers, activists, organizers—who overcame the odds to tell truths that needed to be told. We've made tremendous progress since 1962—dramatically cutting air pollution, cleaning up our water and land, and protecting vulnerable communities from harm—but we still have a long way to go.

You probably know a young woman who is concerned about our climate and trying to figure out her career. Perhaps encourage her to become a scientist or public servant—she could carry on Carson's legacy. Or maybe you know an older woman like my friend Mary Robinson, the former president of Ireland, who could become "an angry granny for climate justice" to do right by her grandkids. Women leaders are more ambitious in their efforts to address climate change than are their male counterparts. The federal government has never acted without a strong push from strong women. You can be one of them.

My advice is to attend a local town meeting, hearing, or event, just as I did when I got my start as a health agent in my hometown of Canton, Massachusetts. I assure you it won't be boring. It's where the messiness, the anxiety, the drama, and the magic of change happen. You will see firsthand the heartbreak of climate change: parents afraid an asthmatic child won't be able to catch their breath, folks living along the coast worrying that their homes won't survive another hurricane season, people wondering if they will have water to drink tomorrow or food to feed their families, seniors stuck in their homes unable to withstand withering heat, and young people who wonder if their future has already been stolen from them.

But you will also see the good things that are happening every day. People working together to invest in clean energy, to build resilience in our cities and coastal communities, to clean up the air, to buy clean cars and demand clean buses for our kids, to rethink our food systems and ensure that no community is left behind. More than 440 cities across the country are committed to upholding the Paris Agreement along with twenty-five states that are home to 55 percent of the U.S. population (if joined together, they would be the third-largest economy in the world). Why? Because people like you and me are showing up in communities, city halls, and statehouses to say, "We demand better."

If you want to make change, if you want your voice to be persuasive and to have real staying power, then get out and see the problems that real people are facing every day in the communities around you. Refuse to sit quietly on issues that impact your world, and be unafraid to lend your voice and share your knowledge when solutions are being debated and decisions are made. Don't let all your energy bounce around on Twitter or Instagram—take it outside, into your community, and let it loose. Follow the young people who are leading in this moment, who are not letting adults off the hook, who are saying, "Enough is enough."

Maybe you're already doing that. Maybe you're already exhausted and overwhelmed by the sheer scale of the challenge and the strength of the forces that are pushing back against you. I get it. I see what's happening politically, and it feels like everything I care about is under attack.

But I refuse to sit around being angry and anxious all the time. I am not going to be a dead woman walking. I am staying hopeful and energized. Why? Because I can. How? I choose to be. Let's turn the chaos of the day into actions that will strengthen our democracy, empower the powerless, and build healthier communities today and a more sustainable tomorrow. Stand tall, stand together, speak up, and get active. You might even need to march to stand up for our democracy, as we did in the sixties and seventies and as we are doing today.

Climate change is not a faraway problem that no one can fix. We have so many solutions already, and innovation can help get us where we need to go faster. Will these solutions fix everything? No. But will they move us forward? Yes. And as solutions grow, they will generate broader engagement. People will see the change and recognize that a low-carbon future is actually good for us. We should want it and we should run toward it like our children's lives depend on it—because they do.

6. FEEL

MADELEINE JUBILEE SAITO

Under the Weather

Ash Sanders

On a bright fall day in 1991, Chris Foster left his differential equations class at the University of California at Davis, bypassed students lounging on the quad, and headed toward the Domes, an on-campus co-op housing development. Although it was November, he was wearing his usual uniform: pink shorts, no shirt, no shoes. At the Domes, he harvested mesquite in a grove of trees and picked wild radishes and mallow in a nearby field. He then walked three miles west to Village Homes, another co-op, which he knew would be a scavenger's cornucopia: full of late-season figs, apples, nuts, and wild grapes. Chris harvested only fallen fruit—he felt this was less invasive than picking from trees, and his aim was to tread lightly on the Earth, to be almost invisible, in order to cause as little harm as possible.

Chris was a philosophy and math major, and he liked to think of himself as the Diogenes of Davis, a reference to the fourth-century B.C.E. Cynic philosopher who renounced wealth and slept outdoors in a large ceramic jar. Chris had made a habit of trying to last the night outside without a sleeping bag. "I couldn't accept the privileges of humanity when I didn't want any part of humanity," he told me. Eating fallen fruit and sleeping outside, however, didn't provide him relief from his feelings of guilt and foreboding. He began to feel a dread that was inescapable and all consuming. A devastating depression that he had suffered a few years before that fall semester returned. Normally a math phenom, Chris started failing his tests. In his apartment, he would sit in the dark—he didn't want to waste electricity—listen to records, and cry. "I felt like I was slowly dying," he said.

A few months later, Chris left Davis to pursue a PhD in philosophy at the University of Kansas. But his condition didn't improve. After having subsisted on scavenged persimmons and radishes for the entire fall term, he'd lost a dangerous amount of weight. His mother paid a visit to campus and, horrified by his appearance, immediately drove him to the grocery store to buy food. At home, Chris's family had a hard time understanding the intensity of the self-denial that governed his life. His father and sister blamed his breakdown on abuse that Chris had suffered as a child; they believed his desire to escape society was a projection, an act of taking responsibility for something that wasn't his fault. But Chris had a different explanation. When he was fifteen, his father had taken him and his sister on a trip to Mount St. Helens. Halfway up the mountain, they had passed clear-cut land. As Chris recalls, one moment there was only evergreen forest and the next moment there was nothing—just bare ground and stumps as far as he could see. A word came to his mind: *evil*. From that day forward, something shifted in him. He didn't want any part of such destruction. By his senior year in college, this conviction had grown into a personal mandate to renounce participation in human society altogether. He was offended by his family's attempts to find explanations in his psychology for problems he thought of as external to him. "Why does my grief have to be because something bad happened to me?" he told me. "They made it sound like I had a psychosis or a mental breakdown and that this is just the form it took, when really, shouldn't anyone who is ethical and compassionate also choose to opt out of this society?"

I met Chris in 2004, when he was a professor at Brigham Young University, more than a decade after he left Davis. He had converted to Mormonism; I had been raised in the church but was on my way out. The first time I saw him, he was giving an animated antihunting lecture, bobbing up and down in front of a whiteboard scribbled with exhortations on animal rights (controversial in an avidly prohunting, conservative state). I was wearing a shirt that said "My name is Ashley Sanders"—something I thought was hilarious, for reasons that now elude me—and when I walked through the door of the classroom, Chris stopped his lecture to say, "Hello, Ashley Sanders." We immediately became friends.

In 2009, discouraged by the failed climate talks in Copenhagen, Chris told me he believed it might already be too late to stop catastrophic climate change. "The old world," he said, "is gone." It was a torch-passing moment. Chris was paralyzed by a conviction in his own failure. He had become complacent, he felt, and addicted to television: the sort of person he used to despise. But I was just beginning my own journey into environmental despair. I was full of guilt and anxiety and anger and fear about a future filled with loss and death. I began to draw my own elliptical lines through the ethics of the climate crisis. I turned off the heat in my house, even during the bitter Utah winters. I was late everywhere, determined to take a bus to another bus to a train. I obsessed over plastic bags and Styrofoam plates and insisted on bringing my own plate to a local sandwich shop. I carried my garbage around for a week and roped my friends into doing so too, each of us hauling a stinking reminder of our consumption from class to class, clearing rooms as we went. I joined a direct-action climate justice group; I planned blockades of city streets and got arrested. I joined with Utah Valley farmers to organize against urban sprawl. I camped in the high deserts of western Utah, trying, with my body and the bodies of others, to physically stop construction on a tar sands mine. I lay down in front of the federal building in Salt Lake City, in the rain, to oppose the Keystone XL pipeline, a poster draped over my body disintegrating in the downpour. I went to Earth First! direct-action trainings, where I learned how to lock myself to fences, pipelines, and other people. When I became devastated by the fact that nothing was happening, I joined radical anticivilization listservs and groups, where people talked seriously about destroying electric, gas, and Internet infrastructure and bringing down industrial society.

I was working fifty-hour weeks, mostly unpaid. My mother, concerned, suggested that I take a break. But I refused. There was no pause button on climate change, so why should I get a break? On some days Salt Lake City, where I lived, had exceptionally bad air quality, a thick soup of pollution settling between the mountains and the valley. The corridor between Salt Lake and Provo, where I'd gone to college, had been completely converted from farmland to strip malls in just ten years. To the south lay one of the biggest open-pit copper mines in the world, to the north was an industrial warren of refineries, and to the west was nuclear waste buried in clay-sealed chambers, reeking of death. That was just the local stuff. Coral reefs were collapsing, ocean ecosystems were overfished, and people in island nations were trapped between salty well water and the swallowing sea.

Meanwhile, everyone around me was fine. Most of them weren't climate deniers, yet none of them seemed disturbed by what they claimed to know. When I talked about how I really felt, environmental leaders cautioned me. "The key is to be positive," they said. "Nobody likes doom and gloom." For me, though, eight years of overwork, stress, and anxiety had taken a toll. My partner and I fought constantly. I had nightmares when I fell asleep and daymares when I read the news. I was sick all the time. I came to hate humankind, its happiness and calm. I went to therapists who stared at me quizzically. I was sad about what? "The end of the world," I said, again and again. Finally, at a loss, they diagnosed me with depression.

Eventually, I left it all. I retreated with my girlfriend to a cabin in upstate New York: no email, no news, no crises. I knew what I wanted. I wanted a world that would last through the century. I wanted a world where my existence didn't mean the end for others. But barring that, I really wanted just one thing: to grieve. To say, *This is unbearable*, and to have people to try to bear it with.

My burnout lasted five years. I still read the news before going to bed—I couldn't help myself. Then, as my partner slept next to me, I'd watch videos of endangered species: a North Atlantic right whale mother swimming with her calves—3 of only 409 left; a free diver swimming with a great white shark, her hand placed gently on its side. I watched as the Shaw Centre for the Salish Sea released a giant Pacific octopus called "the Dude" back into the wild to live his final days at sea. The Dude trundled along the ocean floor, his eight legs whirling. He played with the divers' cameras as they filmed him. Watching these videos felt like a ritual. It was my way of saying goodbye, I suppose, to creatures I had never met but would miss anyway.

The night I watched the Dude's oceanic return, I happened upon an article on the Permian extinction. Two hundred and fifty-two million years ago, the article said, a series of volcanic eruptions spewed plumes of carbon, methane, and sulfur dioxide into the atmosphere, turning the world into a hothouse in a geologic blink of an eye. In response, the vast majority of life on Earth went extinct. It was the largest insect extinction in history. It took the Earth between 8 and 9 million years to fully recover. And then there was the kicker: Now, for the first time in hundreds of millions of years, the article said, we are seeing a trajectory toward Permian levels of carbon in the atmosphere.

So this is it, I thought. *This is how it will end.* I wanted to act. But I was afraid of becoming shrill again, afraid of my urge to grab plastic bags out of people's hands and force my partner to turn down the thermostat and try to fit animal extinction into casual conversation. I was afraid I would become a purist and a martyr, taking all the sorrow and all the suffering on myself because I didn't know how else to reconcile my smallness with the hugeness of the problem.

I wasn't immune to social cues. I knew I was supposed to agree that biodegradable cups were helping or that the empathy of the next generation would pull us out or that not *all* life would be lost—and that humans, resilient, would go on. In each of these conversations, there was a subtext: *This is where you nod and agree. This is where you choose this friendship over party-pooping facts. This is where you stop.*

Sometimes I could do it. Other times I got combative, desperate, contrary. Meanwhile, Chris got married and had two children. When we hung out, he was happier. But he was different too. In his purist days, he'd let his lawn go to seed, refusing to use scarce water resources to keep it green. Now he was living in the suburbs, putting in Kentucky bluegrass. "Why don't you just keep your lawn the way it was?" I said, too urgently. "Because I've been sad my whole life," Chris said, "and sometimes I just want to sit on my green lawn with my wife and feel love." I knew it was just a lawn, but it upset me anyway. Chris had been the one person who understood me, who didn't make me feel crazy or as if I were extreme for what I believed. We argued; I left. I knew I'd hurt Chris's feelings, and I felt terrible for it. I also knew I meant what I'd said. But it went deeper than that. Chris had chosen happiness over constant stress. Was this the ultimate choice: caring for yourself or caring for the world? If it was, I didn't want to know.

Though I quit climate activism for a time, I've kept going to therapy, and I keep confusing my therapists by talking about the end of the world. As it turns out, I'm not alone. A report released in 2012 by the National Wildlife Federation warned that climate change is creating a mental health crisis. The climate scientists, psychologists, and pol-

* icy experts who authored the study estimated that 200 million Americans will suffer from mental illness as a result of natural disasters, droughts, heat waves, and economic downturn. Recent disasters bear this out. In the wake of Hurricane Maria, Puerto Rico's worst natural disaster on record, children developed PTSD at twice the rate of the general population. In the year after Hurricane Katrina, the suicide rate in New Orleans nearly tripled, and the number of instances of depression and PTSD grew to what health experts described as near-epidemic levels. Even people who aren't directly impacted by climate disasters can be affected. According to a 2017 report by the American Psychological Association and ecoAmerica, acknowledging the reality of climate change and its consequences can trigger chronic fear, fatalism, anger, and exhaustion—a condition that psychologists are increasingly referring to as ecoanxiety.

Ecoanxiety can manifest in other serious ways. In 2008, in the midst of a severe drought in Australia, a seventeen-year-old boy refused to drink water because he was afraid that doing so would lead to the deaths of millions of people. Doctors diagnosed him with "climate change delusion" and prescribed antidepressants. When they asked him why he took such drastic action, he said he felt guilty. Since then, his doctors have treated other young people with climate-related psychosis, as well as children with recurring nightmares about climate disaster. In Delhi, a pile of skulls near India's parliament serves as a reminder of the country's more than seventy thousand farmer suicides, which experts say were connected to rising temperatures in the region. (More than ten thousand farmers killed themselves in India in 2018 alone.) Greta Thunberg, a seventeen-year-old Swedish girl who inspired the growing student climate strike movement, says that learning about climate change—and seeing adults' inaction—contributed to a severe depression during which she stopped eating and drinking. Still other activists are turning the violence of climate change on themselves—like David Buckel, a human rights lawyer who in 2018 lit himself on fire in Prospect Park, in Brooklyn, to call attention to the scale of the climate plight.

The emergent understanding of the psychological harm caused by climate change is at the root of a new field known as ecopsychology. According to one of its founders, historian Theodore Roszak, the purpose of the discipline is to define "'sanity' as if the whole world mattered." Ecopsychologists view the Cartesian separation of mind and body—an outlook taken for granted in mainstream medicine—as antiquated and harmful and argue that it can lead to people viewing themselves as separate from the planet they live on. Because traditional psychologists limit their examinations to individuals and their internal maladies, they stamp a "sick" label on patients like Chris and me in an attempt to treat the person instead of treating the problem. Within ecopsychology, the solution is not to pathologize patients but to help them restore their sense of control by reconnecting them with the natural world.

Bruce Levine, a self-described dissident psychologist, believes that forces like climate change have real consequences for the human body. But he thinks it's problematic to consider this sort of anxiety and depression a mental illness. "When you start labeling those problems as a sickness," he said, "political awareness starts to drift away." For example, he told me, PTSD was a diagnosis in vogue after the Vietnam War. The effects of PTSD are real, he said, "but in retrospect, were we not better off calling this problem 'being fucked up by war'?" To Levine, situating sickness only inside the individual is a way for the psychiatric profession to ensure its viability. "It's how we support the power structure," he told me. "We take problems that financial and political assholes created by being uncaring, and now we feel good about solving the problem."

Critics argue that ecopsychology is a flabby discipline without clear boundaries or time-tested methods. They accept recent studies showing that exposure to nature can have positive effects on human health, but they don't necessarily believe the opposite—that the destruction of nature can make people sick. Robert Salo and Joshua Wolf, the doctors who diagnosed the Australian boy with climate psychosis, were careful to note the boy's other symptoms (long-term depression, suicidal thoughts, and hearing voices) and the disproportionate sense of importance he placed on his own actions (believing that his own small water usage would lead to widespread deaths). Other critics have pointed out that climate delusion usually afflicts people who already suffer from other mental health maladies, and that the triggers for psychotic episodes generally take the form of the dominant political or cultural issues of the time, from nuclear holocaust to Cold War–era fears about the spread of communism. These critics might argue that people like Chris and me are predisposed to guilt, compulsion, or a grandiose sense of our own importance based on our past experiences and our chemical makeup, and that if climate change hadn't set us off, something else would have. They might point out that our beliefs endangered ourselves and worried others—so practically speaking, we *were* sick, and we did need traditional treatment.

The critics are right about two things: Chris and I are sick, and we need treatment. But they're missing a critical perspective. If we are sick because our society is sick, shouldn't society be treated alongside the patient? Ultimately, a lot of the disagreement over climate-induced mental illness boils down to vocabulary. We have words to describe the flu or depression or the common cold. We know the contours and symptoms of these illnesses. But when it comes to climate grief, the experience can be hard to define and thus harder to understand and demonstrate. If climate sickness exists in the overlap of the physical and the emotional, we need words for those feelings, a dictionary of sorts that allows us to see patterns in the experiences of individual people. Fortunately, that's exactly what a motley group of philosophers, artists, and doctors are currently working to devise.

In the nineteenth century, the Upper Hunter Valley in New South Wales, Australia, was called the Tuscany of the South for its landscape of rolling hills and winding rivers, trees, and savannah. In the eighties, when Glenn Albrecht arrived there to write a book about two ornithologists, he was confronted with an entirely different vista: a massive clear-cut coal mine. He returned regularly to the valley over the years, and by the first decade of this century it had been gutted, with more than 16 percent of the valley floor exposed. The homes in the path of the mines had been bought and bulldozed; those remaining, on the edge of the pits, were covered in soot. Dust and grime hung in the air. Klieg lights lit up the sky twenty-four hours a day. Birdsong had been replaced by the noise of detonations. It felt, Albrecht said, "like walking into an apocalypse."

Between visits, Albrecht returned to the University of Newcastle, where he was an environmental studies professor. Even among the town's sun-soaked hills and beaches, he couldn't shake the feeling of distress that the mine had provoked in him. Eventually, after he'd made many trips to the Upper Hunter Valley, locals began calling him. Their homes were being destroyed by the mines, they said, and they asked for his help. Albrecht was an environmental writer and advocate, and he was accustomed to people asking for his advice, but the desperation, sorrow, and panic in these voices constituted something different: palpable grief. Locals already felt that the mines were making them physically sick, with asthma and cancer, and led to high rates of birth defects. They were also angry that the mines destroyed their houses. But it was apparent to Albrecht that the residents' pain was emotional as well as physical. It seemed they were not losing just their homes but their *home*, their sense of belonging to a place. He started to see a relationship between environmental and psychological distress, between the health of the land and the health of its people.

As a philosopher, Albrecht determined to identify and name this condition. The word he came up with was solastalgia, a portmanteau of the Latin *solus*, which means "abandonment and loneliness," and "nostalgia." "Nostalgia" has not always had the warm and fuzzy connotations it does now. When it was first used, in the seventeenth century, it described a diagnosable illness that afflicted people who were far from home but could not return. Soldiers were particularly susceptible, as were people forced into migration by conflict, colonization, and slavery. The cure, it was thought, was simply to return home and be soothed by familiarity—otherwise the sufferer's distress would continue, even to the point of death. To Albrecht, if nostalgia was a sickness caused by the displacement brought about by seventeenth-century globalization, solastalgia was its twenty-first-century counterpart.

Once he had a name for the condition, Albrecht set out to record its existence. In April 2003 he returned to the Upper Hunter Valley with a team of social scientists. He conducted interviews with residents, asking them about their families, their houses, and their history in the valley. Meanwhile, his research partner, Nick Higginbotham, conducted statistical research. Higginbotham used the Environmental Distress Scale to compare the health of Hunter Valley residents to that of a control community that had not been exposed to mining. He mailed out surveys to residents in both communities asking them to rate the frequency of environmental disturbances they had witnessed near their homes; how threatened they felt by them; the emotional, economic, and physical impacts they felt; the actions they'd taken to address the situation; how much they trusted government and industry information; and the degree to which loss of place, or solastalgia, had affected them. People living in the hardest-hit mining areas reported more environmental disturbances, higher levels of threat, and greater emotional and physical effects than the control group. Most important, though, they reported acute feelings of solastalgia, agreeing strongly with phrases like "I miss having the sense of peace and quiet I once enjoyed in this place" or "I am ashamed of the way this area looks now." After nearly two years of exhaustive qualitative and quantitative research, Albrecht and his team were able to define solastalgia as not just a word or a concept but an empirical reality.

Albrecht's study on solastalgia was the first of its kind. Since then, additional research has demonstrated a connection between environmental and psychological distress. Meanwhile, the word has traveled far; Albrecht has sparked something of a movement in art, politics, and philosophy, as more and more people search for a way to describe their feelings of grief and loss about climate change and environmental destruction. In October 2018, when I spoke to Albrecht, he was in Sweden, talking with the Sami about the effects of forest fires on their land and psyches. Albrecht worries about his fame, however. "The fact that the word is gaining traction means we're in deep shit," he told me. "I want it removed from the English language as quickly as possible."

Lise Van Susteren, a practicing clinical psychiatrist in Washington, DC, has developed her own words for climate grief. Van Susteren started reading about the health consequences of climate change in the early aughts. After *An Inconvenient Truth* came out, she signed up to be a climate educator through Al Gore's Climate Reality Project. Her scientific training gave her valuable insight into the consequences of climate change, but it also led her to place too much faith in data. "I was unrealistic about the human mind and our capacity to deny," she told me over the phone. "I honestly thought: We'll look at where we are, we will see we have to get carbon emissions down to three hundred fifty parts per million." She led classes and waited for change to happen—but things only got worse. "We were not only not bringing numbers down—they were accelerating," she said.

Van Susteren started having trouble sleeping. After getting into bed and closing her eyes, she would be ambushed by intrusive images. She would see refugees surrounded by barbed wire, animals trapped in the path of a hurricane, people stranded in floodwaters. The worst image was of a child. It wasn't any child she knew, but a sort of representative for all children. The child looked at Van Susteren and asked the same question again and again: "Why didn't you do anything?"

As a psychiatrist, Van Susteren recognized her symptoms. The stress, the insomnia, the intrusive thoughts—they read like PTSD. And yet the trauma she was imagining hadn't happened yet, or at least it hadn't happened to her. How could she feel trauma before the cause? But her stress was undeniable, and it was debilitating. "You can breeze through the newspaper and try not to read about insect Armageddon, the glacial ice sheets melting, wildfires, climate refugees, and storms," she said, "but it registers on your psyche. Whether you like it or not, it does." Van Susteren coined a new term for her condition: pre-traumatic stress disorder.

I have had symptoms of pre-TSD for much of my adult life. When Van Susteren told me about the disorder, I thought of the night I learned about the Permian extinction and the argument I'd had with Chris about his lawn. At the time, I'd just felt frustrated and confused. Now I had a word for these feelings. The sleepless nights, the obsessive YouTube-ing, the tendency to place huge importance on everyday decisions—they were rooted in a fear of the future. I was picturing all the small decisions across the world and multiplying them by time. I was grieving the future in the present, and—since there were so few people who wanted to talk about it—I was grieving it alone.

If we accept the emergence of solastalgia, pre-traumatic stress disorder, and other climate-induced ills, what's to be done about them? How do we confront the reality of climate change and convince others to do the same? The environmentalist Alan AtKisson calls this predicament "Cassandra's Dilemma," after the princess of Troy who appears in Aeschylus's tragedy *Agamemnon*. Cassandra is blessed with seeing the future, but her gift is accompanied by a counterbalancing curse: No one believes her prophecies. AtKisson connects the myth to climate action: The more a person knows about environmental destruction, the more they will try to warn others, and the more others will, in fear and defensiveness, resist them.

The environmental movement has been collectively trying to solve this problem for years. Its solution has been to gather more data. It hasn't worked. In the wake of environmental inaction, many activists have started to shift the emphasis toward emotions, not facts. A key strategy is to name those emotions and normalize them.

To find out more about this approach, I visited the Bureau of Linguistical Reality, in San Francisco, in 2018. The bureau is tucked into a corner of the Yerba Buena Center for the Arts. At the time of my visit, elsewhere in the city, Indigenous activists were blocking the streets in protest of California governor Jerry Brown's climate policies. I was there during the Global Climate Action Summit, a convergence of world leaders, politicians, and environmentalist bigwigs determined to compensate for Trump's withdrawal from the Paris Agreement. In a corner of the arts center, behind a small table, I found Alicia Escott and Heidi Quante, the two founders of the bureau. They were dressed in military greens covered in patches of the starburst etched into Carl Sagan's Voyager Golden Record. Next to the table were posters, propped on easels, that read: "shadowtime," "gwilt," "casaperdida."

The mission of the bureau is to create a dictionary for the Anthropocene—not a dictionary of scientific terms but a lexicon of words to describe the destabilizing experience of living through mass climate change. Escott and Quante conceived of the project during an unusually sunny San Francisco winter, when they confessed to each other that they felt joy at the warm weather (they could wear sundresses) but guilty about what it meant. They wanted a word for that feeling, something that captured their ambivalence. They came up with "psychic corpus dissonance." Other words include "ennuipocalypse": the idea that the end of the world might be not a Hollywood Armageddon but mundane and almost normal; and "NonnaPaura," the desire to have children or grandchildren, mixed with a fear about the world they'll inherit. The bureau's mission combines philosophy (because words forge the way we see the world) and politics (because new words create new possibilities for action). But it's also absurd: the military uniforms, the game of making up new words to describe highly specific feelings, the free patches they give out. They look like experts from another planet, reporting on their interactions with sentient life on Earth. The strangeness invites curiosity, said Quante. Instead of throwing data at people, the bureau invites the public to reflect, feel, and collaborate. "I noticed that the approaches of the environmental groups weren't effective with people who didn't already agree with them," she said. "So obviously talking at people and telling them what to do doesn't work."

When I walked in, I noticed that Escott and Quante had placed a sign near the museum entrance announcing "Word needed," followed by a definition: "The misdiagnosis that an individual's depression or mental illness is derived from something wrong with them personally, when the depression may actually be induced by living in a society that is ill or broken." The bureau doesn't usually provide a definition in need of a word, but so many people had talked to them about this feeling that they'd decided to crowdsource it.

The definition immediately made me think of Chris. Escott and Quante related to it too. After the failed Copenhagen climate talks in 2009, Escott hid out in her studio in the Castro, broke and devastated, and started a strange experiment: She wrote letters to extinct animals like the desert rat kangaroo, which disappeared from Australia in 1994, and filled the animals in on what they had missed in the intervening years. Quante told me that one of her earliest memories was learning that so many things around her were alive—the trees, the grass, the frogs. It terrified her to realize the harm she was capable of. One day, after it had rained, her mother made her walk along a worm-strewn sidewalk, and she screamed as she was dragged along. "We're killing them!" she said. "We're killing them!"

When Quante debuted the bureau at the Paris climate meetings, a woman from Haiti approached her, almost in tears. The fact that other people were talking about feeling grief, the woman said, made her feel less isolated and crazy. An environmental lawyer came by later to look for a place to sit. When he learned what the bureau does, he confessed to Quante that his job weighed on him heavily. Every day, he told her, he went to court to fight against environmental destruction, but as a lawyer, his only tools of expression were the languages of rationality and law. He could never say what he actually felt. As he spoke to Quante, he began to cry. "No one has ever asked me what I felt about this before," he said.

"There's such a suppression," Quante told me. "Our artwork is showing that almost everyone is experiencing solastalgia. A lot of people are depressed. They're not alone." Quante believes that this shift to people realizing they are part of a greater collective can lead to social change.

Throughout the day I visited the bureau, people walked up to the table with looks of excitement or relief. "I've got a word for that depression thing," said a young woman. "How about 'sociopression'?" It was clunky (many of the bureau's words are), but Escott and Quante were game to discuss it. The woman, an actor, told them that her job was to make people feel things. Why, she wondered, would we stamp out feelings about *the* most devastating problem? "It's like people create all these diagnoses to diminish the larger existential reality." She paused, thinking about what she had just said. "What about 'distrance'?" she said. "Like a mix of 'distance' and 'distraction'?"

The rest of the day brought requests for a word describing the jaded feeling of constantly receiving bad news; a word for experiencing generations of environmental trauma; a word for the way that ash and smoke from a forest fire turn the sky purple and the trees a bluish gray. (Two months later, when California was engulfed by forest fires, the bureau was creating another word, "brokenrecordrecordbreaking," for the feeling of déjà vu experienced when reading that this year's catastrophe records are, again, the highest ever.) One woman approached the table uncertainly. She had a bristle of gray hair and Hello Kitty tattoos on her arms, and she wanted to suggest the word "eco-pooper." "You know," she said, "like that person who brings up some depressing environmental thing every time you talk?" Then she turned to me and asked what word I would invent. I wondered if I should tell her that *I* was an eco-pooper: her worst nightmare. Instead I said, "Maybe something about repressing the desire to talk about climate change at parties." She looked at me and shrugged. "Well, yeah, no one likes that. Like, *The world is ending. Pass the cheese puffs!*"

Today Chris lives in a gingerbread-style house in downtown Salt Lake City with his wife and two kids. It's a life his Davis self never could have imagined. But although his life may look normal, he's still devastated by the state of the planet. When I called him in late 2018, he was stewing. He'd just been to Utah Valley University, in Orem, to see a mutual friend of ours named Ben present his latest climate research. After the panel, while they were talking about some of the grim news, Ben declared that he considered himself an optimist, and that if Chris kept talking pessimistically, things would never change. "He didn't say I was factually wrong," Chris said. "Just that I felt wrong." Chris likened optimism to a boat headed toward the edge of Niagara Falls. "It's not the guy who says we're fine that changes the boat's course," he said. "It's the guy who says we're all going to die."

Chris acknowledged that this quality doesn't make him easy to be around. In 2018 he informed his ethics class that the amount of carbon in the atmosphere was over 400 parts per million, and that there was no hope of getting it down to 350. In the back of the class, a student started crying. "If I didn't have hope, how could I live?" she asked. Chris had an urge to say: *Exactly*. But he felt bad about making her cry.

Chris's depression sometimes makes his life feel unbearable, but attempts to feel better don't sit well with him either. He keeps busy writing papers, attending activist events, staging protests. He often feels like a failure. He suffers from nearly constant intrusive thoughts. When he goes to bed, he sees the faces of billions of animals—bears, mountain lions, factory-farmed cows and pigs—all suffering, all staring at him. He tries to imagine what extinction must be like. "There's starvation involved and there's horror and betrayal and there's no recourse," he told me. "You just have to go on until you die. Until you're all dead."

After I hung up the phone, I thought about Chris for days. I thought about his feeling of never being or doing enough—I both admired and worried about it. I thought about his insistence on pessimism in the face of the pressure to be optimistic. Research favors the optimists, I had to admit. Most humans respond better to manageable, hopeful amounts of information tied to concrete, doable actions. But I wondered if, in the end, what is important isn't choosing optimism or pessimism but honesty with oneself. It seemed to me that for his entire life Chris has battled for the space to vent his rage and anger and sorrow, to be given a place to grieve and people to grieve with. And this, more than anything else, is what I needed in him and what he needed in me.

I thought about this again one bitter day in January 2019, while lying on my back on the Rockefeller Center skating rink in New York City. That morning, I'd met up with a dozen activists from a climate group called Extinction Rebellion, and we'd laced up our skates, held hands, and wobbled onto the rink. At exactly 2:30 P.M., we dropped to the ice and formed the image of an hourglass. Above us, another protester climbed from the mezzanine onto the giant golden Prometheus statue and hung a banner from its arms reading "Climate Change = Mass Murder." On Fifth Avenue, more protesters blocked traffic; others staged "die-ins" representing the victims of the climate crisis. After five years of paralysis, I was back. I was still afraid: afraid of more burnout, afraid of caring, afraid it wasn't enough. But I'd come because I'd read the group's principles. Number one: Tell the truth.

We lay on the ice for half an hour while tourists from Texas, Florida, and California skated around us. As I shivered, I thought about climate grief. Ever since I'd learned about climate change, I'd wanted to believe that people could stop it. In order to stop it, I knew we'd have to be honest with ourselves about it—to have the courage to face its true size and consequences. But all the advice I'd ever gotten, all the social cues, all the role modeling, had told me to lie and pretend. For years I had wanted to find a place where I could be honest, because I wanted to believe that we—humans, people, I—might still be able to act meaningfully in the face of our own extinction.

Around the time of the protest, I told Chris about the bureau's dictionary for the Anthropocene and asked what word he would add. Chris teaches philosophy of language, and I knew he'd be into it. But he got even more excited than I'd expected, rattling off a long list of possibilities. Ultimately, he settled on one to describe both others and himself. He called it "ignore-ance," or "returning from a state of consciousness to a willed state of not knowing." That's where he was now, he said, and where so many people insist on being. He was surviving, but he didn't admire himself. "You do it by pretending," he said, as if teaching me how. "You pretend that this life is okay, that college football is fun, that driving is normal. You pretend, to justify living a lie."

Chris feels daily that he's not doing enough. In reality, he is accomplishing more than almost anyone I know. But he's had to make certain concessions in order to stay alive. He's happier now that he's not sleeping in the hills outside Davis, subsisting on boiled wheat and self-loathing, but that also means he spends more time in a place of willed not-knowing. In the language of Bruce Levine, the dissident psychologist, Chris is sick not because he is inherently anxious or depressed but because he lives in a sick society in which both belonging and survival depend on inaction.

Maybe the word we need is not one for a sickness. Maybe we need a word for a difficult truth: that when the world is ending, our health depends on closing ourselves off to awareness of this fact. Where you choose to draw your boundaries is arbitrary, not rational. If you draw them wide—if you include trees and refugees and animals and whole nations—you will be sick from overwhelm and will be seen as crazy. But if you draw them narrowly, you'll suppress more and admire yourself less—which is its own sort of sickness.

I decided to ask Chris. Was he sick? He thought about it for a long time. "I don't know," he said finally. "But I know this: If your heart is breaking, you're on my team."

Our feelings are our most genuine paths to knowledge.

—Audre Lorde

Mothering in an Age of Extinction

Amy Westervelt

I am sitting in a hotel room in DC, still in bed under the covers at noon, unable to move until I absolutely have to for a work thing, cycling through the catalog of losses: sea turtles, because of the sand getting too hot; most birds, forced to shift their habitats and migratory patterns to find suitable food or temperatures, are fucked; coral, bleached and drab; California . . . I probably shouldn't live there for much longer. Then I get a call from my husband and the kids: "Hi, Mom. I miss you!" the littlest one says. "Mom, why do you have to go away for work?" his older brother chimes in.

Work trips are usually a slight respite from the carrying of immediate and long-term concerns at the same time, all the time. At home, it's relentless: My three-year-old tells me with absolute glee that he can't wait to be four, and I think about how much less habitable the Earth is getting every year; my seven-year-old asks for a snack and then peppers me with questions about what college will be like, and I think of the Greenland ice sheet melting and wonder, with a lump in my throat, if college will even be a thing when he's older, while also being careful not to give him too many junky snacks.

I am worrying about their present and their future, whether I'm doing enough for them and enough for humans, and it fills me with both grief and white-hot rage. Parents have always worried about their kids' futures and presents at the same time, and that goes double for mothers who are marginalized in any additional way. Wearing climate goggles is a new version of this special fear, performing hope when you feel terror, preparing your kids for the worst without letting on too much. Trying to make them resilient but not bitter, prepared but not terrified.

Lately I've had a lot of conversations with my friends about their decisions to have or not have kids. Mostly they fall into the latter camp. Often because of climate change. They either don't want to add to it or wouldn't want to subject their kids to a future that has already been darkened for them. In either case, as my friend Anna

Jane put it recently, "It pisses me off that climate change even comes into the decision making around that at all."

Mom friends, meanwhile, share their climate grief with me, their full-body anxiety. The dread they feel for their kids' futures, a weird and profoundly sad thing for a mother to feel, looking into the face of a cheery five-year-old. The low-level panic that hums through their bodies while doing normal everyday things like packing lunches and planning play dates, the sudden realization, as they're stressing about balancing work and parenting, that things are about to get so much worse as "work-life balance" morphs into "work-survival balance."

Sometimes I wish I'd made a different decision on the kid front. I've been a climate reporter for twenty years; I knew the score. What was I thinking? But it's also mostly the images of my kids (and yours) embroiled in resource wars, losing their innocence far too young, wearing gas masks to high school (because of the fires, duh), and an accompanying sense of righteous indignation on their behalf that propels my work on the subject.

If that conjures up images of Sarah Palin and her "hockey mom" shtick, I get it. So-called maternalism has had a long and fraught history. Maternal rhetoric can feel really gross and gendered and filled with expectations forced on female bodies.

But it can also be quite powerful. Mothers have always been key organizers and social-justice activists. Maternal rhetoric was a cornerstone of the civil rights movement, the abolitionist movement, and, more recently, the gun-reform movement. What feminist scholar Patricia Hill Collins calls "othermothering" and other researchers refer to as community mothering has been a powerful part of organizing in Black communities. The idea is that you nurture not only your own children or family members but also the community around you—it isn't done exclusively by women, although it often has been. Anthropologists like Sarah Blaffer Hrdy have seen this approach across multiple communities, and in fact Hrdy credits it with the human brain evolving toward emotional intelligence.

In many marginalized communities, "community mothers" are the ones leading the charge to clean up the water, get transit working, hold police accountable, and protect and care for their neighbors—

actions that regularly lead to major statewide and even nationwide shifts. So much so that Collins rails at some of the critiques mainstream feminists have leveled at maternal activism. "This type of thinking sets up a hierarchy of feminisms, assigns the type engaged in by U.S. Black women and women in Africa a secondary status, and fails to recognize motherhood as a symbol of power," Collins writes in *Black Feminist Thought*. "Instead, the activist mothering associated with Black women's community work becomes portrayed as a 'politically immature' vehicle claimed by women who fail to develop a so-called radical analysis of the family as the site of oppression similar to that advanced within Western feminism."

Second-wave White feminists often rejected motherhood as a patriarchal institution that oppressed women; it was a key part of the rift between White women and women of color in the feminist movement. The civil rights movement, on the other hand, embraced and harnessed the maternal; organizers there understood that while modern society may use the role of mother to punish women for seeking independence, that was not driven by motherhood but by patriarchy—and capitalism, which desperately needs care work to remain free labor. Perhaps this is yet another lesson the climate movement could and should learn from the civil rights movement. In addition to the ideas that you don't need eternal optimism and hope to fight for what's right, that knowledge and information rarely shift power structures, and that community organizing is a critical (maybe *the* critical) part of addressing intractable social problems, civil rights organizers understood that mobilizing maternal activism, tapping into the ethic of community mothering, is a powerful tool in the organizing toolbox. It's one that's been completely underutilized in climate. Mostly because anything other than hard data and charts has been avoided by the movement for decades.

But as we collectively begin to acknowledge climate grief and the need to process it in order to act (thanks in no small part to the work of climate-psychology researchers like Renee Lertzman), mothers have often been left out of the conversation. We talk about population, about whether it is or isn't "responsible" to have children in a climate-changing world—that conversation comes up every decade or so—but we rarely hear about how today's mothers are processing climate grief for two (or more) or how our panic might be directed toward action. We talk about the youth climate activists, but we rarely hear from the parents who are enabling, and inspiring, their activism, fueled by their own desperation to protect their kids from the worst-case scenario. On climate, for the most part, mothers are a wasted resource, and we can't afford to waste anything anymore.

I spend a lot of my time reading the letters of dead men. Men who felt some kind of way about climate change—or "the greenhouse effect" and "global warming," as it was popularly known, before another guy convinced oil companies and politicians to change it to "climate change," which sounded suitably vague and natural. Letters from scientists sounding the alarm about a planetary catastrophe headed our way and letters from businessmen warning of a profit catastrophe. "Scientific evidence remains inconclusive as to whether human activities affect the global climate," then–Exxon CEO Lee Raymond said in a speech he gave to the American Petroleum Institute's annual gathering in 1996, nearly twenty years after his own company's scientists warned that "man can afford [a] five-to-ten-year time window to establish what must be done. It is premature to limit use of fossil fuels, *but they should not be encouraged.*" (Emphasis mine.)

"Everyone agrees that burning fossil fuels releases carbon dioxide and that such concentrations in the atmosphere are rising," Raymond said. "But it's a long and dangerous leap to conclude that we should therefore cut fossil fuel use. . . . I'm not proposing we dismiss the possibility of climate change. But I am asking that we apply common sense to the debate. Many scientists agree there's ample time to better understand climate systems and consider policy options. So there's simply no reason to take drastic action now." This motherfucker.

These speeches and memos infuriate me. And again I'm reminded of parenthood and its requirement of endless trade-offs, thousands of choices between short-term and long-term benefits. Do I leave my kid sleeping during an unexpected nap and take advantage of a free hour of work time but deal with the consequences later when he's up until ten, or wake him up now and deal with his tantrum but get him to bed at a normal time? Do I let the other one stay home from school when he really doesn't feel like going, or insist that he go? Opt out of the work-work-work American thing, or enforce the rules and make sure he can function in the society we live in today? Should I just let him play Roblox for an extra hour on the iPad so my husband and I can have an uninterrupted conversation, or get him off that thing because it turns him into a creep? It's a constant choice between me, my kids, and the greater good. And I almost never feel like I'm making the right decision.

As the climate changes rapidly around us, the decision making just gets harder. What will actually help them most in the years to come? How do I prepare them for a society that might operate by completely different principles?

Meanwhile, fossil fuel execs are like those shitty parents who always choose their own convenience and then expect everyone else to accommodate their asshole kid. They have to balance the present and the future, themselves and everyone else, in daily decisions too, and they choose short-term gain for themselves every day. And then tell us that we need to make more responsible consumption decisions; that's the real problem. As I write this, Big Oil, which speaks about "simply supplying a demand," has moved into plastic, looking for a new market for its oversupply of natural gas and even crude oil. The industry has even decided to get into the manufacturing of plastic itself, in facilities that belch carbon dioxide.

Here, too, there are parallels. Especially for moms who work in climate. Every day I have to choose between what's best for my own kids—probably more undistracted time with their mother—and what's best for everyone's kids—doing everything I can to ensure a livable planet. What's best for me and what I feel compelled to do for my family and my community and my species and my planet. I spend most of my days looking for ways to tell the story of climate change and why it's more of a threat now than ever, why action is so delayed; hoping to help people understand, process, and get to a point of action; thinking if we can just mobilize at a massive scale in the next few years, maybe, just maybe, we can avoid global chaos. And for a lot of those hours when I'm worried about humanity, I am shushing or ignoring the little humans who are in my direct charge.

Just before Thanksgiving last year, I had to evacuate my kids during a California wildfire. Rain was predicted the next day, which would help to put out the fire, but I worried that if the storm was too intense, it would kick off mudslides. *This is climate change*, I thought to myself. Then I was summoned urgently to the bathroom to wipe my three-year-old's butt. I was panicked, because I grew up in California; I know how quickly fires can move and how little notice they give. But I had to pretend for the kids that it was no big deal.

I've always thought community mothering was a cool concept and have yammered on at more than one person about how it's the approach we should all be taking to both raising children and being responsible members of society. But I'd never really internalized it, never considered "community mother" as a role I myself might play, and I'd still always felt enormous guilt over missing time with my kids, even if I was working on something that felt truly important for the public in general. The increasing urgency of climate change has made it all click.

In a documentary about Dolores Huerta, the legendary organizer who cofounded the United Farm Workers, her kids are very open about the fact that there were times when they were growing up that their mom chose her cause over her children. There's resentment there, but there's also pride. They talk about how they had to share her with the world and how as they got older they realized that her work was important to their whole community. Working on climate accountability has helped me crack the work-life-parenting balance: I don't sweat the small stuff, appreciate every minute I spend with my kids, make a point to hold and cherish these moments before the storm, and see my work for the greater good not as a conflict with motherhood but as an integral part of it.

Anthropocene Pastoral

Catherine Pierce

In the beginning, the ending was beautiful.
Early spring everywhere, the trees furred
pink and white, lawns the sharp green
that meant *new*. The sky so blue it looked
manufactured. Robins. We'd heard
the cherry blossoms wouldn't blossom
this year, but what was one epic blooming
when even the desert was an explosion
of verbena? When bobcats slinked through
primroses. When coyotes slept deep in orange
poppies. One New Year's Day we woke
to daffodils, wisteria, onion grass wafting
through the open windows. Near the end,
we were eyeletted. We were cottoned.
We were sundressed and barefoot. *At least*
it's starting gentle, we said. An absurd comfort,
we knew, a placebo. But we were built like that.
Built to say *at least*. Built to reach for the heat
of skin on skin even when we were already hot,
built to love the purpling desert in the twilight,
built to marvel over the pink bursting dogwoods,
to hold tight to every pleasure even as we
rocked together toward the graying, even as
we held each other, warmth to warmth,
and said *sorry, I'm sorry, I'm so sorry* while petals
sifted softly to the ground all around us.

Loving a Vanishing World

Emily N. Johnston

Even just breathing by the ocean, I feel different. The smell of the sea is home for me, the brush of the waves on the shore, the spark and flutter of sun on the water like innumerable languid butterflies.

At a beach in British Columbia's Gulf Islands last year—on my first real vacation in almost three years—I felt much of the loosening that I often feel at the coast. But I've known for a long time that humans and other species are in profound trouble and that the seas are rising. I've known for a long time how much is at risk, and I'd gone to BC to have the time to think and write about those risks and how we can move forward in a way that matters.

So sitting there, on sand and the countless soft shards left behind by clams and mussels and oysters over decades, I couldn't unknow the fact that the oceans are beginning to die. A single one of the several ocean garbage patches contains nearly two trillion pieces of plastic. There are microplastics in sea salt worldwide. Fish populations are collapsing. Whales and dolphins are suffering profoundly from the din of the sonar used by oil companies and the navy. Seawater is acidifying so fast in the Salish Sea that oysters are struggling to build shells. And perhaps most troubling of all, phytoplankton levels have fallen rapidly since 1950—and phytoplankton is not only the base of the marine food chain; it also produces about half of the world's oxygen. This last fact, by itself, should be enough to make us address the intersecting crises in the natural world immediately.

The truth is that the ocean that looks so beautiful and unchanging is on its way to becoming a vast garbage dump full of plastic and heavy metals, where jellyfish thrive and other life plummets. It will not smell the same. Its colors will change. And as ocean ecosystems falter, seabirds, of course, will die too.

It's a constant question for me every time I'm entranced by the beauty of this world: What does it mean to love this place? What does it mean to love anyone or anything in a world whose vanishing is accelerating, perhaps beyond our capacity to save the things that we love most?

The stakes are unnervingly clear if we look at the Earth's five previous extinctions—particularly the end-Permian, in which most of life on Earth was wiped out. That extinction was triggered by greenhouse gases from the Siberian Traps, which led to temperature rise and climate destabilization. This happened extremely suddenly in geologic terms—but with temperatures and greenhouse gas concentrations that were rising orders of magnitude *more slowly* than we're causing them to rise now.

So it's not just our grandkids; it's not just low-lying or hot and dry places; it's not just humans; it's not just orcas or the Great Barrier Reef or monarch butterflies; it's not even "just" the oceans (upon which so many species, including people, depend). What's at risk now, as best we can tell, is the better part of life itself.

The potential loss of so *much* life is clarifying, because there is only one medicine for any of it—for any of *us*—and that is the restoration of a thriving natural world, beginning with the near-term end of fossil fuel use. If we make real progress, we can almost certainly tip the balance for *some* individuals and species—at least for a while. And that's a good thing: To help some people live longer lives with some stability is much better than not to do so, even if it doesn't last for millennia. To save some species is far better than to save none. What could be a more meaningful way to spend our lives?

The word "sacrament" comes from the Latin word for "solemn oath"—used by early Christians, interestingly, as the translation of the Greek word for "mystery." This work is, in the deepest sense, both a solemn oath and a mystery; it is a sacrament. We are walking into great darkness, and the light that guides us must come from within.

Would you risk your life for someone you love? Would you work day in and day out to give someone a chance at a decent life?

Then you know what to do—the imperative of it, if not yet the details.

It feels important to say here that it's not reasonable to expect to feel *hopeful*, at least not very often—but our gift, and our task, is far more powerful than sunny feelings, because we still have the chance to *make the space* for hope—to act in such a way that hope might exist for others who come after us.

Not everyone can focus on this work, of course—many people are too full up with the difficulties of their daily lives; they have small children or elderly parents who need care; they work sixty hours just to keep food on the table or commute three hours a day because housing near their jobs is too expensive. But if you can, then the world needs you, and it needs you right now, because anything that we do this year or next is worth ten of the same thing ten years from now. Nothing has ever been more important than holding the world well back from any of the ecological tipping points that we haven't yet crossed, and some are perilously close.

This makes us, however unintuitively, the most powerful people who have ever lived. It's easy to see that this is true in ways that have been devastating for Earth. Nothing has threatened our ecologies more than the extraction and burning of fossil fuels and the affluent consumer culture that corporations have persuaded us is normal: the flying and driving, the eating of meat, the buying of things that we don't need made of materials that nature has no experience breaking down, and all of this made devastatingly easy by the fact that those of us who do most of this live far from the places where our pollution poisons the air and water, where animals are slaughtered, where plastics are made or discarded.

But we have also been granted an astonishingly beautiful gift that has never before been given to humans: the chance to shepherd human and animal life into the coming centuries and millennia, when we know that much of it would otherwise disappear. That's a power that should make us very humble and a privilege that can motivate us profoundly. In a way, our darkness—the knowledge that without our great effort, many or most of Earth's creatures will vanish—is what reveals the light within, the seed of life and possibility that we share with all of Earth's life, the one that we can carry forward. For better and for worse, we are the ones at the intersection of knowledge and agency.

Faced with the certainty of devastating loss, I sometimes struggle to breathe. But the incredibly freeing truth is that life on Earth isn't concerned with my sorrows, and those who are already struggling to save their kids' lives or their homes aren't interested in whether I'm grieving or uninspired. They need me to *do* something.

What we can do bears some relationship to how we feel, of course—but it's not dictated by it. I can organize people to go to a hearing to testify for new public transit or against a new pipeline, even if I feel demoralized—and doing so will likely make me *less* demoralized. I can make calls to elected officials, even if I suspect it's not going to matter. And I can organize my community to support carbon-absorbing forestry and agriculture, even if I don't really believe it will be widespread enough to make a difference.

Here are two truths: To some of us, much of the time, it feels exceedingly unlikely that humans will survive this—yet it's a simple fact that if we respond robustly, we *can* survive this. Despair is an accurate reflection of the peril we face, but it isn't a predictor of the future; it's devastatingly nearsighted. To see beyond what despair sees—to move from the feeling toward the possibility—calls for things we have in abundance: love, imagination, and a willingness to simply tend the world as best we can, without guarantee of success. Any one of these can get us started.

Everybody has different skills and different temperaments. I'm an introvert, so organizing didn't come naturally to me—and still doesn't, really—but I've learned ways I can be effective by leaning on other people and letting them lean on me. We can best utilize our abilities, in other words, by valuing and supporting the abilities of others. At my activist group, 350 Seattle, we have one volunteer who spends a day every week doing our books, another who does all the tricky work on our database, another who writes all our thank-you notes. We even have a retired massage therapist who offers us free massages.

All of the work is critical in this moment, and we must do it with humility, learning as we go, taking on both the deeply satisfying tasks and the unpleasant or routine ones. We don't have to believe they're adequate—we only have to understand that *not* doing them would mean we'd decided not to care for this world and ceded the greatest power we've ever had, in order to . . . what? Watch television? Do yoga? Make really cool apps? How will those choices feel when you're dying and you know the world is too? How do they feel even now, when whole towns go up in flames, when "hundred-year" storms happen with sickening frequency?

Imagine if even 10 percent of the country started engaging deeply, even just one day a week. Our possibilities would be—will be—entirely different from what they are now, because our existing system, the one that's hurtling us toward disaster, depends entirely upon our disengagement from one another. It depends on our belief that we can't really have an impact on what we care about—that we should all instead simply focus on our own personal fulfillment: a thing that isn't even possible in the absence of deep engagement, with the natural world as well as with one another.

Our emotions matter to *us*, of course, even if not to those who are suffering the most. Neither being sunk in despair nor becoming automatons will help us do this great work with love, and thrive, which is what will allow us to do all we can. But much of the time, we can experience our emotions and still do the work we need to do. Feelings need not be destiny.

When my mother was dying, my grief was profound, and occasionally I avoided calling her because of it; such complicated feelings didn't fit easily into a difficult and busy time in my life. But I called the next day. And the next, and the next. And I traveled to see her, though not as often as I would have liked to. And I knit her an alpaca blanket—in meetings and on car trips and whenever else I didn't need my hands.

We can feel fear and grief and anger—we can even feel avoidant sometimes—and still attend to the world's very real and immediate needs. And in truth, serving the world's needs is the only thing that I have seen consistently *lighten* that fear and grief and anger in others, and the only thing that has done so consistently in my own life. There is also, perhaps oddly, joy in this work. It's made me more deeply alive and connected, with a clearer perspective on what matters, and has surrounded me with friends who share my care for the world.

In any moment, we can choose to show up.

I've been thinking lately about the "campsite rule," by which we're supposed to leave a place or a person in at least as good shape as we found them. As a species, this is currently impossible—if you multiply even a single plastic bottle by nearly eight billion people, it's clear that we've been devastating for the web of life. But not evenly, not by any means. The great majority of our billions have done extremely little harm, and what harm they have done was outside their control—they cut trees to cook with wood because it was all they had, or they commuted to work in a particularly polluting car because it was what they could afford and there was no good public transportation.

Here in the United States, any adult of working class or above (anyone who flies or commutes in a car or lives in a single-family home) could spend the rest of her life planting trees and taking plastic out of the ocean, and as an individual, she would still be far, far in the red to the living world: Her campsite would have a stream poisoned by mountaintop removal or fracking, and it would be an antibiotic-resistant mound of plastic and animal remains. Individual actions can slow the growth of that mound, but they're simply not enough to make a dent in it.

We did not intend this harm, but we have partaken of it; given the reign of neoliberalism and the lies of the fossil fuel industry, living as social beings almost required that we partake. But now that same social nature requires something new of us.

We can't undo what we've done simply by being nice and Earth-friendly people—nor even by dying. There is too much we need to heal, and we have to change the path that we're on. We have beautiful work to do before we die.

In 2015 I was part of the #ShellNo fight against Arctic drilling. I had held a vigil with a friend almost three years before, when Shell's rigs had also been in town. But this time we were part of a much larger group, and experienced organizers came to town, young women who knew so much more than any of us did that it was truly humbling. I threw myself into the campaign and was aided by the fact that the same week that Seattle's public port announced it would be welcoming the Polar Pioneer rig to Elliott Bay, the journal *Nature* outlined the energy projects we *could not engage in* and still hope to avoid truly catastrophic climate change. Arctic drilling was one of these projects. So I pointed out the juxtaposition to every port commissioner, news crew, and journalist who would listen: We've just been told this is a civilization-threatening project, and *the City of Seattle is enabling it.* About a dozen of us made up the core group—organizing people to go to the hearings, learn kayak safety in advance of our on-water blockade, shut down the port by land for two days, and generally put our outrage to good use. For about six months, I lived and breathed this fight.

I did all of this purely because it was the only way I could look myself in the mirror. I didn't have any faith that it would "work" in any clear or immediate sense, and I knew that wasn't quite the point. Still, when the other "kayaktivists" and I delayed the departure of the rig for only about an hour, my heart sank. All of that work and love—for what? I knew that it mattered that we'd drawn an international spotlight to the dangers of Arctic drilling . . . but it was hard to feel like that counted for much as I watched the rig leave for the fragile Chukchi Sea.

I cheered up notably several weeks later, when friends in Portland stopped one of the associated vessels—without which drilling couldn't start—for a full forty hours. They hung from a bridge with streamers in the blazing heat—the most beautiful and effective small-group action I'd ever seen, supported again by scores of people in boats.

But the real transformation in my perspective came at the end of September, when Shell announced that it was abandoning its Arctic drilling quest. Publicly Shell cited operational issues, but a company source told *The Guardian* that, also, they'd been greatly surprised by all the protest and made acutely aware of the risks to their reputation. By which they meant: us. A project they'd sunk years and billions of dollars into, in remote waters far from population centers, had ground to a halt because of a dozen or two core people in Seattle and Portland and a thousand or so others who participated once or twice or all the way through.

That's leverage. Archimedes knew: Give us a long enough lever, and we can move the world.

And what it *felt* like, for a month or two, is that I personally had stopped Arctic drilling. We had slain a dragon, and for a little while I felt invincible, a thing I had never felt before. Not in an egotistical way, really—I knew very well how much I'd needed the wisdom, skill, and numbers of the others; I had, after all, organized a perfectly useless vigil two years before. But it helped me to understand my power—a power rooted in working with others so that we might be greater than the sum of our abilities. A power rooted in showing up and being willing to learn. Working with this small group of people, I had scored a meaningful win for ecological stability by using leverage: a dozen folks in one circle focused more or less exclusively on this, supporting another few dozen in the next concentric circle out, giving a day or two a week, supporting another couple hundred showing up every few weeks or months, and several hundred more who probably came out only once or twice but who thereby made it clear that this was something that many people *cared about*.

We haven't had such a magical or consequential win since then. We *have* had wins, though, which are often—like that one—simply the absence of devastating losses; it's why an energy insider referred to the Pacific Northwest as the place "where energy projects go to die."

We bought a tiny bit of time. Perhaps we saved a few humble species—for a few decades or centuries—by making one of the worst projects just a little more expensive and shifting public opinion just that bit more toward what might be, a couple of years from now, some legislation ambitious enough to finally matter. I know for sure that we shifted some people's faith in the usefulness of protest.

All that is very good, but it's not enough: We need you. This fight is just flat-out *too big* for the existing folks to do what needs to be done. We are exhausted, and we need you.

So what *does* it mean to love the world (or anybody or anything in it), and how we can think about love and hope and imagination, even when the coming decades start looking like the end of the world? Because they will. In places like New South Wales, Australia, or Paradise, California, they already have.

Practical "what you can do about climate change" articles are important, but if we can't metabolize new ideas about hope and meaning, then lists of ways to engage will fall radically short when we're grieving—and we'll lose our chance to find or make possibility *within* that grief.

It was *so beautiful* up in British Columbia last year. At one level of my soul, I can simply appreciate it—we may as well, after all, and loving this vanishing world feels like a kind of prayer sometimes. At another level, I am overwhelmed by the grief of knowing what's coming—everything from the vast human suffering to the very specific and local near-certain loss of the seventy-three remaining Southern Resident orcas—creatures so extraordinary and intelligent that I feel privileged simply to share a bioregion with them and ashamed that I cannot save them. But at the deepest level, I've begun to invert time and shift metaphors so that I can see not only loss but gain. A world with millions of people versus one with none, or a world where half of extant species survive versus one with, say, 5 percent—these are worlds absolutely worth fighting for—though from this relatively full moment in time, it's hard to celebrate those millions or that half, knowing what will have gone missing.

We have to love not just this vanishing world, in other words, but the many worlds we can still prevent from vanishing.

When we think about accelerating extinction, it's like looking at the terrifying narrows of an hourglass, where only a few will slide through. So sometimes I imagine myself instead on the far side of something more like an ecological birth canal. How many of Earth's beauties can we help to survive the passage into the next era? Is each one not a gift we can safeguard to the world by our actions?

Isn't that who we want to be?

By fighting for social change, we aim at what the world can be—and help other people see both the problem and the possibilities. This is also the power that Greta Thunberg and other youth climate strikers have; it's utterly self-evident to them that we must all change our lives to respond to this crisis, and the purity of their understanding is transmitted to us.

In the world of your imagining, can you see children living who might have drowned, stable communities that might have burned, species hanging on—and understand those as things to fight for, for the rest of our lives? We can rejoin the web of life. We do not have to be its destroyer. But our last best chance is now, and countless tasks lie ahead of us.

So some night soon, when you're on your way home and are tired and unsettled and thinking about all that you have to do; or next week, when this plea has faded away and someone asks you to do something you're not sure you want to do; or better yet, in a few weeks, when you realize that there's something *you* can do by bringing a group of people together, remember: In any moment, we can choose to show up.

We can let them kill this beautiful world—or we can get to the beautiful work of making space for a decent future.

Being Human

NAIMA PENNIMAN

I wonder if the sun debates dawn
some mornings
not wanting to rise
out of bed
from under the down-feather horizon

if the sky grows tired
of being everywhere at once
adapting to the mood
swings of the weather

if clouds drift off
trying to hold themselves together
make deals with gravity
to loiter a little longer

I wonder if rain is scared
of falling
if it has trouble
letting go

if snowflakes get sick
of being perfect all the time
each one
trying to be one-of-a-kind

I wonder if stars wish
upon themselves before they die
if they need to teach their young
how to shine

I wonder if shadows long
to just-for-once feel the sun
if they get lost in the shuffle
not knowing where they're from

I wonder if sunrise
and sunset
respect each other
even though they've never met

if volcanoes get stressed
if storms have regrets
if compost believes in life
after death

I wonder if breath ever thinks of suicide
if the wind just wants to sit
still sometimes
and watch the world pass by

if smoke was born
knowing how to rise
if rainbows get shy backstage
not sure if their colors match right

I wonder if lightning sets an alarm clock
to know when to crack
if rivers ever stop
and think of turning back

if streams meet the wrong sea
and their whole lives run off-track
I wonder if the snow
wants to be black

if the soil thinks she's too dark
if butterflies want to cover up their marks
if rocks are self-conscious of their weight
if mountains are insecure of their strength

I wonder if waves get discouraged
crawling up the sand
only to be pulled back again
to where they began

if land feels stepped upon
if sand feels insignificant
if trees need to question their lovers
to know where they stand

if branches waver at the crossroads
unsure of which way to grow
if the leaves understand they're replaceable
and still dance when the wind blows

I wonder
where the moon goes
when she is in hiding
I want to find her there

and watch the ocean
spin from a distance
listen to her
stir in her sleep

effort give way to existence

The Adaptive Mind

Susanne C. Moser

Put your ear to the ground and you can hear something strange happening—in college classrooms and at farmers' markets, in church basements and conference side events, in office storage rooms–turned–secret employee gathering spaces, in private social media groups, in bookstores and libraries.

Be still and listen: There are silent tears flowing, people talking in hushed voices, trying to contain their anger, fears, grief, and despair. At first you will have to listen hard for the sounds of these times, but they get louder as you pick up their vibrations: *We are hurting. We are afraid. We are exhausted.* At least some of us. Some of the time. And some of us again and again. And then you will know: *You are not alone.*

More and more I find myself in such spaces. One Sunday afternoon, I went to a local church to join a "Climate Listening Circle" organized by members of Extinction Rebellion and Climate Action Now. Some twenty people—almost all of them women, twenty-somethings to seventy-somethings—gathered to share feeling vulnerable, worried, overwhelmed, and in despair. They came seeking a place to be all of themselves. They came to do not just *climate* change work but, as one put it astutely, *culture* change work.

As we sat in a circle, people spoke to what it means to be human in the midst of an ecological and climate emergency. One brought a giant blue papier-mâché ear to symbolize her (and our) readiness to listen and witness. Everyone was acutely aware of what's wrong with our planet, our government, our economic models, and our mental models. In the end, they left knowing they have sisters (and brothers) with them in their internal and external struggles.

Showing Up for the Coming Darkness

I've been to many such circles—participating in or leading them. As a social scientist eager to have my finger on the pulse of society, I come not just to participate but to see how people are learning to show up for the emotional onslaught of climate change, how people self-organize for mutual support through this century of transformation.

Most, like this one, bring together climate activists. Rarely, if ever, among them are there climate scientists or so-called adaptation professionals—such as resource managers, infrastructure engineers, or city planners—charged with helping communities prepare for and manage the impacts of climate change. In this circle, I was once again intensely aware that those practitioners were not in the room. Scientific reputation, professional decorum, and the ever-present expectation of expert impartiality and the absence of outward expression of emotion—all of them stand in the way of these public figures joining such a circle. Where, I wondered, do *they* go?

Over the last few years, I have been thinking a lot about just that: What do humans need to get through this megacrisis? I don't mean which clean technologies, hard-hitting carbon prices, or incisive policies. I mean: What do the *people* working on climate change need in order to keep showing up for the long and psychologically taxing journey ahead?

After all, every day the problem of accelerating climate change becomes more apparent through new scientific reports, news headlines, and the consequences manifesting all around us. There is a growing recognition now that humanity is headed into a world in which three kinds of change prevail:

> *Ongoing and accelerating change*, deviating from the familiar patterns of seasonal cycles and long-term stability, confronting us with pervasive uncertainty, surprises, and often outright not-knowing.
>
> *Traumatic change*, brought on by catastrophic events like Hurricane Maria devastating Puerto Rico, Hurricane Harvey flooding Houston, or the Camp Fire ravaging Paradise, California.
>
> *Transformative change*, a deep, fundamental change to the systems in which we live. Transformative change could come through large-scale technological interventions, economic reconfigurations, fundamental policy changes, sociocultural shifts, complete

makeovers of how people earn their livelihood, and relocations from areas becoming uninhabitable.

Few, if any, in government, academia, or other institutions have been formally trained in how to deal with these massive and simultaneous challenges. And yet unprecedented levels of change circumscribe the outlook for public officials, community leaders, urban planners, engineers, emergency first responders, health professionals, and decision makers everywhere.

All who—by virtue of passionate outrage or their day jobs—are called to lead and support their communities in this future need insights, training, and support to keep showing up for the darkness looming before us.

Fostering the Adaptive Mind

What climate professionals need in order to keep showing up I call the "adaptive mind." It may be something we all need. The adaptive mind comprises a set of propensities, capacities, and skills that allow an individual—embedded in social networks and institutions, as we all are—to respond with agility, creativity, resolve, and resilience to the kinds of stressful changes described above.

That's still vague, but it's a start. It's at least a map of where to look to begin building our individual and collective capacity for psychosocial resilience.

In 2017 my colleagues and I launched a project called "The Adaptive Mind." To understand what these propensities, capacities, and skills might be, we are drawing on every relevant discipline and area of practice, including psychology, military training, business management, education, sustainability, organizational development, ecosystem management, leadership studies, arts and the imagination, social capital theory, faith and spirituality, and emergency response. We are exploring which of them are learnable, and we will work with psychologists and networks of climate practitioners to deliver trainings, offer resources, and build peer groups to support their strenuous day-to-day work.

Without such support, the risk of burnout is great, as is the likelihood of losing highly skilled professionals just when their skills are needed most.

Already, according to a first-of-its-kind survey of adaptation professionals we're conducting, four out of five respondents say they feel burned out. The perfect recipe of attitudes and feelings underlie that finding: Respondents expressed a dogged commitment to helping their communities prepare for the coming changes and to finding feasible, equitable solutions, yet at the end of every day, they go home feeling their work is never enough. How can they ever take a day off to recharge their batteries? How can they find a sense of equilibrium, even if fleeting, that allows them to return to work fortified?

And isn't this how so many feel? That in the face of an overwhelming problem, doing nothing is not an option? In fact, for many, doing *something* is the very antidote to despair. Doing something can give us hope and purpose, while sitting still puts us at risk of falling into a hole of darkness. And when we take moments of stillness to consider where society stands, the outlook seems dark, and only getting darker.

Understanding climate science means knowing that worse is yet to come, even if society were to finally accelerate its commitment to greenhouse gas reductions. It means letting the implications of that science for our communities sink in. It means, every day, staring in the face the many hurdles to implementing real change on the ground—political, institutional, financial, and social.

Plus, there are social injustices at the root of these challenges that pervade the impacts as well as the responses to them. Add to that deeply entangled legacies of structural racism and the fact that most adaptation professionals are White and privileged. So the work of addressing climate change is inseparable from the work of healing injustice and dismantling racism.

Learning how to stand in the heat of these struggles requires self-transformation and transformative leadership—fundamental components of the adaptive mind.

Raising Recognition of Psychological Distress

Bringing attention to the psychosocial challenges of working as a climate professional is not to elicit sympathies, nor to wag a finger about ignoring a growing problem. It's certainly not intended to dissuade anyone from this career path. Far from it.

The point is simply this: Burnt-out people are less effective people. Burnt-out people can become sick people. Burnt-out people leave their jobs and are replaced by less experienced people. Burnt-out people aren't equipped to serve a burning planet.

Ensuring that those helping communities through this difficult time of change have all the help they need to be high-caliber leaders is enlightened self-interest. Some employers will simply see it as a matter of economic savings (Gallup found that burnout costs American society dearly every year). Lawyers might see it as a matter of avoiding expensive liabilities. Others might simply think of it as a matter of kindness and, ultimately, as a question of what kind of society to live in: whether it's one in which the golden rule applies or one in which nobody cares about another's well-being.

We need wise, effective, and resilient people helping communities prepare for and navigate this transformative time.

To be sure, for more than a decade, psychologists and psychiatrists have been raising the alarm about the coming wave of psychological distress due to the climate crisis. Nearly every article and report available recognizes the potential traumatic effects of natural disasters on those directly or indirectly affected. They warn of immediate impacts of shock, grief, fear, and PTSD, but also of secondary psychological problems such as increases in alcohol use, domestic violence, sexual abuse, etc. Some also discuss increases in violence and suicidality due to extreme heat. But only just recently have psychologists begun to acknowledge phenomena like climate grief, ecoanxiety, and anticipatory or *pre-traumatic* stress disorder as pervasive emotional experiences that refer to the whole overwhelming phenomenon. Some have proposed classifying the entirety of climate change as a "trauma."

And while these challenges have been recognized as problems for the general public, very little attention has been paid to the particular stresses experienced by those who work on behalf of the general public on climate change solutions every day, with two exceptions. First, there is an emerging recognition of the emotional labor done by those expected to be unemotional and objective—namely climate scientists. Second, some have also written about the burnout common among advocates and activists, as well as the emotional burden laid on youth.

With the exception of first responders such as emergency workers and firefighters, the psychological strain on the broader category of climate professionals has not been adequately recognized. A lucky few find good individual counseling services. Apart from that, there are scarce relevant resources or support systems to deal with the emotional toll that climate change is exerting on these professionals.

Taking Care of Ourselves

Clearly, such professional support will be crucial, and it should be developed collaboratively by climate experts, psychologists, and those affected, so as to meet needs adequately. But until these support systems are in place, what can be done? Lots.

Know you're not "crazy"; you're human. Climate change not only wreaks havoc with the physical world but also affects people psychologically. Knowing what the psychological and mental health impacts of climate change might be, what ecoanxiety, climate grief, and trauma are, can be reassuring and empowering and helpful in simply not feeling "crazy." You, we, are *not*!

Take care of yourself. It sounds boring and obvious, but it bears repeating: Psychological self-care is not a luxury. A stressed-out body and mind work less efficiently and effectively, get impatient, grumpy, even aggressive, and—if the stress is chronic—can suffer negative health consequences. By contrast, a balanced, resilient mind is a kinder and more compassionate, alert, productive, and effective mind.

Take a break and enjoy simple pleasures. Pause your activism, work, emails, news feeds. Take some time to breathe, stroll at lunch, turn off your phone/apps/TV, buy or pick yourself flowers, go to the movies, get a massage, or simply do something completely different.

Do what brings you immediate joy. Maybe chase a ball around with your dog, go to game night, or have a solo dance party. Perhaps pick up a neglected hobby like photography, gardening, or bowling.

Maintain healthy routines. This could include getting enough sleep, working out, spending time with family and friends, and preparing and eating delicious, healthy food. Consider practicing yoga, qigong, or meditation.

Do some mental hygiene. Check on your habits of thinking, such as catastrophizing, looking only at the bad news, insisting on despair and helplessness, or judging your and others' actions as inadequate. Balance action with reflection to notice what is working and not working. Maintain a focus on progress, milestones, and positive changes. Cultivate grounded hope.

Focus your attention to selected things, not everything. See how you can (with colleagues, family, friends) prioritize, share, and delegate tasks/responsibilities. Let go of the least important things.

Enhance your emotional coping capacity. Get in touch with where you're feeling things in your body. Allow yourself—alone and in a safe setting—to name and express your feelings. Learn to witness your emotional ups and downs. Reconnect with your own kindness and compassion. Know that many have very similar experiences and you are not alone.

Nourish yourself spiritually. Engage in rituals of healing. Participate in religious or spiritual communities that are meaningful to you. Spend time in nature. Practice mindfulness, prayer, singing.

Seek out social support. Hang out with people who are fun, give you a sense of being cared for, and actually support you. Seek out a mentor. Spend social time in your community of choice. Share your concerns with trusted friends or family.

Get professional support. While the majority of psychologists aren't yet fully aware of the climate reality and its psychological implications, nor of the challenges faced by those working on climate change every day, there are now efforts under way to grow a cadre of climate-conscious psychologists. Climate professionals shouldn't have to rediscover what is already well established about self-care, trauma-informed services, or capacities for deep change. And psychologists should seek out climate professionals too, to learn about the topic and what those in distress most need.

Help change institutional cultures. This might entail establishing supportive well-being services and policies, fostering self-care and not rewarding overwork, providing training and retreat opportunities to staff, and establishing expectations of skillfulness in interactions with distressed colleagues and stakeholders. Overall, foster an organizational culture that acknowledges the need for restorative rest, shared leadership, and mutual accountability so we can keep showing up for the work ahead.

Showing Up as an Act of Rebellion

Dealing with the causes and impacts of climate change is work for the long haul. Confounding challenges—such as personal crises, other environmental issues, persistent social injustices—make bearing witness to climate change harder some days, just as little joys and victories in other parts of life can make it easier to show up again for the climate crisis.

To support that hard work, my hope and commitment is to help ensure that those doing it have the psychological skills and capacities, as well as the peer and institutional support, to effectively and compassionately face the challenges ahead. In such a world of accelerating change, the well-being of our hearts and souls must be reestablished to their rightful place as relevant, essential. This is work that will resurrect what we have—wrongfully—abandoned for far too long, bring back the feminine in all of us, and restore our human wholeness.

Even though transformation at organizational and institutional levels will take time to come into place, we can start supporting ourselves and one another right now. So maybe look for giant ears set up at libraries and farmers' markets. Look for gatherings in church basements and café corners. Take time out whenever you need to, as the journey is long and burnout is painful. If there can be "Fridays for Future," why not also have "Somedays for Sanity"?

Caring for our hearts and minds, rejuvenating our bodies, reconnecting with one another, and deepening into our deepest purpose is taking our psyches seriously. This is an act of rebellion against the extinction of soul. That, too, is climate change work. It is culture change work. And it is essential.

Home Is Always Worth It

Mary Annaïse Heglar

I washed up on the streets of New York City in 2006, fresh out of college and my southern roots fresh on my voice. I'd worked hard for my bachelor's degree in English, and the only sensible thing to do seemed to be to take it to work in either publishing or journalism. Those industries lived in New York, and everyone told me they were dying, that New York was too expensive now and I'd never make it. But these were the same people who'd told me that it was fine to rack up student debt at a hopelessly expensive, elite liberal arts institution because I was sure to land a great job and pay it all back, easy peasy. If they were wrong then, they were probably wrong now. I liked those odds.

Journalism was hard to crack into, but publishing was easier. To keep my options open, I did both. By day, I worked on publications at a university. By night, I volunteered with a local lefty newspaper, making futile attempts to fit my voice into a mold as a "real journalist." I never succeeded—a fact I am now very happy about.

It was there, in that volunteer newsroom, that I met what I have come to not-so-affectionately know as a "doomer dude." The major news outlets were still largely covering their ears and mouths when it came to "global warming," as it was then derisively, controversially, dubiously known. But my little paper, *The Indypendent*, bravely decided to break the silence by dedicating its entire June issue to the crisis on the horizon.

Each month, we had elaborate open-floor editorial meetings to plan the upcoming issue. This one drew a particular strain of peculiar volunteers out of the woodwork. As the unmistakable scent of aluminum-free deodorant wafted through the air, I found myself surrounded by tall White men with remarkable sunburns and disheveled hair and cargo pants. They towered over me with their tales of woe. "There's really no point anymore. Humans are done for!" Their voices swelled with glee. Perhaps they saw the horror on my face; one of them hovered down to my eye level and offered, as if it were a consolation, "Oh, don't worry! The Earth will be fine! She just needs to get rid of *us*!"

Their wistfulness, their cavalier acceptance of not only their own demise but also mine and everyone else's, was as perplexing as it was intimidating. If the Earth needed to get rid of them? *Sure*, I thought, *that makes sense. But what did I do?*

I'd been in New York almost a year by this point, but I was still too young and too southern to know my way out of a mansplaining vortex. I nodded and smiled, then cried the whole way home.

They were significantly older than me and didn't seem to register (or even consider) how many of my dreams they were crushing. According to them, I was not entering adulthood; I was entering a furnace. Almost on accident, their joyful nihilism effectively placed environmentalism on a shelf that was too high for me to reach. I, in turn, limited myself to the issues on the shelves within my reach: police violence, income and education inequality, homelessness, etc. Fixing what I could while the world burned.

I didn't know how to speak up then. I didn't know how to tell them that I couldn't bring myself to give up on myself before I'd even begun. But I've grown up.

The Chickens Have Roosted

I joined the climate justice movement in earnest in 2014. Since then, I've come across a good many doomer dudes. They have books. They host panels. They are prolific tweeters. They are legion. And they're almost always White men, because only White men can afford to be lazy enough to quit . . . *on themselves.*

I've given them a new name: "de-nihilists." They've absorbed all the disaster porn and memorized all the worst-case scenarios, and they can rattle them off with rapturous pleasure. They've not only accepted that our fate is sealed, they've found comfort in it—so certain are they that they know how this unwritten story ends. I can only assume they're very obnoxious in movie theaters, with their constant attempts to spoil the ending.

To an extent, I get it. There's no denying the severity of our crisis, at least not anymore. There's no more putting it off on "future generations." No more "stopping" global warming. It is here. The chickens have roosted.

Perhaps the most terrifying part of it all is the uncertainty. The not knowing. It's unmooring, unnerving. I can understand reaching for something, anything, to be sure about. And if you've nursed at the teat of toxic masculinity, I can understand why you might think you need to be the one in the room with clear-eyed vision. The one who knows what's coming next. But no one does.

Because the thing about warming—whether we're talking about the globe or a fever—is that it happens in degrees. That means that every slice of a degree matters. And right now that means everything we do matters. We quite literally have no time for nihilism.

Hope Doesn't Spring Eternal

On the other hand, to be fair, the climate community can be maddeningly hawkish about *their* narrative and *their* messaging. *We must be hopeful! We can't be alarmist! We must adhere to strict scientific nuance at the expense of clarity and urgency and beauty! We must leave no scientific rabbit hole unexplored! Nuance. Nuance. Nuance.*

This sort of tone policing makes the climate conversation impossible to have with any real honesty. Not in a world where what we used to know as "potential impacts of global warming" now have proper names. Theoretical and hypothetical climate models have come to life in the forms of Hurricanes Maria and Dorian, Typhoon Yutu, Cyclone Idai, the Camp Fire, and bushfires that have sprung up so quickly we can barely name them, so we just call them "Australia."

We've entered an age where these tragedies fade and blend into a continuum that we struggle to recognize as "normal." Meanwhile, the climate community's insistence on hope everlasting begins to sound anything but realistic. At best, it is emotionally stunted. At worst, it's downright sociopathic.

Not to mention that in order to have this type of hope, you have to be able to identify and articulate the solutions that would justify it. And that favors a certain type of expertise and raises the price of admission to the climate conversation to an astonishing, astronomical, unforgivable rate. We can't afford all these gates and all these gatekeepers. Again, we don't have time.

Granted, this reflexiveness is a by-product of decades of relentless, bad-faith attacks from both industry and government, but the result is the same. It's exhausting. It's ineffective. And it's alienating. Honestly, it's not terribly unlike the de-nihilist narrative. Both are mansplaining paradises. Both smack of the privilege wrought from the deluded belief that this world has ever been perfect and that, therefore, an imperfect version of it is not worth saving or fighting for. Both represent two sides of an overprivileged pendulum swung too far.

What a Wonderful, Imperfect World

But the community that prides itself on its scientific nuance can learn to embrace emotional nuance. It is absolutely possible to prepare for the disasters already, terrifyingly, upon us while also doing our damnedest to quit baking more in. We can acknowledge the storm of emotions that comes with watching our world burn. We can process those emotions and pick ourselves up to put the blaze out as best we can.

Because it's worth it. Because we're worth it.

We don't have to be Pollyanna-ish or fatalistic. We can just be human. We can be messy, imperfect, contradictory, broken. We can learn the difference between hopelessness and helplessness. Because what if we've been doing the equation backward? What if hope isn't what leads to action? What if courage leads to action and hope is what comes next?

I've never seen a perfect world. I never will. But I know that a world warmed by 2 degrees Celsius is far preferable to one warmed by 3 degrees, or 6. And that I'm willing to fight for it with everything I have, because it *is* everything I have. I don't need a guarantee of success before I risk everything to save the things, the people, the places that I love. Before I try to save myself.

Even if I can save only a sliver of what is precious to me, that will be my sliver and I will cherish it. If I can salvage just one blade of grass, I will do it. I will make a world out of it. And I will live in it and for it.

We don't know how this movie is going to end, because we're in the writers' room *right now*. We're making the decisions *right now*. Walking out is not an option. We don't get to give up.

This planet is the only home we'll ever have. There's no place like it. And home is always, always, always worth it.

7. NOURISH

MADELEINE JUBILEE SAITO

Solutions Underfoot

Jane Zelikova

As a kid growing up in Ukraine, I loved playing in the dirt. Using only a spoon as a shovel, my older brother told me, if I dug long enough, I could get to America. My family eventually immigrated here, by way of planes, trains, and automobiles. In college I took an ecology class to meet a science requirement and discovered that playing in dirt could be a career. It didn't take much convincing to change my major from anthropology and Russian literature to ecology: The labs were outside and we had the chance to dig with much larger spoons.

I loved the feeling of soil in my hands and sought out research projects that gave me the chance to get dirt under my nails, to color my fingerprints with earthy hues. As a budding ecologist, I studied forest regeneration, counting and measuring tree seedlings and saplings in the Georgia pines and later in Costa Rica. In graduate school, I turned my attention to small but mighty organisms: ants. My first summer of fieldwork was spent crawling around on the forest floor painting unsuspecting winnow ants with fluorescent paint. I needed to track them as they traveled through the forest litter so I could find out what happened to the seeds they dispersed.

It turned out that winnow ants were moving northward and up in elevation to escape the more frequent and oppressive heat brought by climate change, effectively abandoning the plants that rely on them for seed dispersal. The relationship between winnow ants and plants is a seed-dispersal mutualism, one where the plants are more dependent on the ants than the other way around. Rachel Carson said that "in nature, nothing exists alone." Ecology is all about relationships. Organisms exist in relation to one another and to their environment, shaping conditions of life and being shaped by them. And ecology, like all relationships, is messy. In a rapidly changing climate, ecology is also a study of consequences. As I tracked fluorescent-hued ants in the Georgia heat, I realized that the ecological relationship between ants and the plants they disperse, which took thousands of years to develop, was coming apart in a matter of decades. And that is how I found my way, inadvertently, into climate change research.

For more than twenty years, I have spent my waking hours doing experiments that simulate climate-model predictions and measuring how plants, critters, and microbes respond. I diligently observe and record the responses of these organisms, all the while living inside a global, man-made climate experiment. As a scientist, I am both fascinated and shocked by how rapidly climate change is disrupting ecological relationships that evolved over millennia. But science also reminds me that we are not powerless. We can slow and perhaps even halt climate change with some human ingenuity and courage, and some help from the invisible multitude under our soil-covered shoes. We can help save ourselves, and the planet we depend on, if we're willing to play in the dirt.

Soil has been described as the skin of the Earth. Soils are vast ecosystems that support all agriculture and every single terrestrial biome. Without Earth's exceedingly thin layer of soil, humanity as we know it would not exist.

Over the last twelve thousand years, we have lost about 133 billion metric tons of carbon from this soil, stripped away as humans converted native grasslands and forests into agricultural fields and rangelands, roads and cities. The main driver of this loss—the plow—revolutionized farming and fundamentally altered the trajectory of human history. Hunter-gatherer nomads gave way to farmers. Once farmers could grow enough food to feed more than their immediate families, they freed up the rest of the community to specialize in other trades.

Today we use more than a third of the planet's land to grow food for 7.8 billion people worldwide. Our production of food and fiber spans about fifty million square kilometers, an area equivalent to Russia, Canada, the United States, and China combined. The United Kingdom may be a few decades away from losing all its soil fertility, and other parts of the world, including Brazil, central African countries, and parts of Southeast Asia, are in a similar predicament. We may have close to ten billion people living on Earth by midcentury. Ensuring food for all is going to put a lot more pressure on a precious, disappearing, and resoundingly underappreciated resource: soil.

There is no technical innovation that can bring soil back—no machine that can do what thousands of years of rock weathering and biological activity have achieved. Rebuilding soil requires fundamentally rethinking our reliance on technology and chemicals to deliver what soils can do on their own: support life.

To bring back soil, we have to feed the microbes.

A tablespoon of soil contains billions of microbes. These tiny bacteria, fungi, protists, and archaea make up the bulk of life in soils. There may be a trillion species of microbes on Earth—99.999 percent still undiscovered. Though invisible to the naked eye, microbes collectively hold more carbon than all animals combined. Billions of tons of carbon sit underground, three times more than in the atmosphere.

The Dutch scientist Antoni van Leeuwenhoek first saw tiny organisms, or "wee beasties," as he called them, in a drop of pond water in the 1670s. With the rapid innovation in molecular and computational tools, we are finally getting a peek into their lives 350 years later. We still have very little idea who they are or how they make their way in the world, but we know that this unseen majority drives an invisible engine that hums under our feet—in our yards, in parks, on our farms and ranches, in natural and engineered ecosystems across the world.

Microbes are the movers and shakers of carbon sequestration. They transform organic matter from plants and animals into soil organic carbon (SOC) and other nutrients, a process that builds soil fertility and draws down carbon from the atmosphere and locks it away.

How do microbes do their alchemy? To understand microbes, you have to understand plants.

From giant sequoias to microscopic cyanobacteria, from lush coastal rainforests to arid rangelands, photosynthetic organisms drive the single most important chemical transformation of carbon on our planet. Plants are proverbial straws that draw in carbon dioxide from the air through photosynthesis. Combined with water and sunshine, plants transform carbon dioxide into carbon building blocks. With each sunrise, plants kick photosynthesis into high gear, and as the sun sets, carbon drawdown ceases until dawn. With the advent of new instruments, we can now observe the Earth's daily gas-exchange cycle in real time, the culmination of billions of individuals that exhale carbon dioxide and all the plants across the world that absorb it to power photosynthesis.

Plants don't use every ounce of carbon they capture through photosynthesis for their own growth. The extra seeps out of roots into soil as exudates, a secretion hungry soil microbes consume. Microbes use some of the exudates for their own growth, keeping that carbon underground, and release some back into the atmosphere. Plant roots not only directly feed carbon to microbes but also stimulate the creation of pores, small pockets between soil particles and clumps. Scientists have only recently uncovered how important soil pores are for carbon storage. Locking away carbon in pores or in clusters of soil particles protects it from being consumed and partially exhaled by microbes. It turns out that the bulk of stored carbon is actually dead microbes, known as microbial necromass.

Having more carbon in the soil is transformative. It means better water infiltration and higher nutrient and water retention. As soil health improves, agricultural fields become more resilient to the ups and downs of a changing climate. That resilience helps plants grow more consistently, even when the weather is fickle. To put it simply: Healthy soils matter now more than ever.

There are tried-and-true ways to get more carbon into our soil, like reducing tillage (aka less plowing), planting a diversity of crops, and growing cover crops during the fallow seasons to keep the soil covered. These practices ensure living roots in the ground year-round, delivering a fully stocked carbon buffet for the microbes and powering the microbial engine humming underneath our feet. On farms and ranches, this means farmers can become less reliant on costly external inputs like fertilizer and irrigation. Healthier, carbon-rich agricultural soils support more robust and sustainable agricultural operations—a win for the people who grow our food and a win for climate.

Diversity is the magical elixir for better soils. It took over two hundred years of research and debate for scientists to generally agree about the importance of diversity, but now it's finally clear. Diverse plant communities deliver a greater variety of carbon exudates through their roots, feeding a rich patchwork of microbes. The virtuous cycle continues as microbes transform and store more carbon, delivering additional soil-health benefits that help support diverse and thriving plant communities.

Though microbes operate at microscopic scales, they're a huge deal. The "wee beasties" hold the power to restore our diminished soils and work in collaboration with plants to absorb and store atmospheric carbon for centuries or millennia. That means soils can deliver a simple but major climate solution: They can help us put carbon back where it belongs. The key is shifting how we grow our food so we can reverse the loss of soil and harness the magic of photosynthesis and microbes. Their powers combined could draw down about 10 percent of the carbon dioxide we emit every year!

And the people best poised to bring this climate solution to fruition are the same people who steward our land and put food on our tables. Farmers and ranchers can lead the transformation in how we grow our food, with support and collaboration from nature's carbon cycle.

We have a long way to go to address the challenges in front of us and make the most of soil's carbon-banking potential. Luckily, we're not starting from scratch. Scientists have the tools to track changes in soil carbon over time, and we are getting more familiar with the secret lives of microbes and their carbon-transformation powers. There is a growing community of farmers and ranchers who are excited to regenerate their land. The agricultural solutions are not new and don't hinge on technological breakthroughs. Building better soils and drawing down carbon is ready for prime time and can be implemented on millions of acres across the country and world tomorrow. Visionary climate leaders can help make this a reality.

For too long, we've had a small, homogenous chorus of climate heroes who have regularly overlooked solutions that are literally underfoot. Despite touting the benefits of diversity, the climate movement has been largely affluent and White. As a scientist, I know that diversity is the main ingredient for hardy and productive ecological systems. Why would it be any less vital in addressing climate? When it comes to climate solutions, diversity of perspectives and approaches should be foundational; we need it to foster ingenuity and resilience. Without diversity, innovation is stale, and we end up with "change" that looks like the status quo.

While the old guard pushes the same ways of thinking that got us into the climate crisis, impactful leadership is already happening silently, microscopically. Microbes are helping create and sustain ecological relationships that form the foundation of thriving soil ecosystems and a sustainable food system. As entrepreneurs and investors dream up huge machines that can pull carbon out of the air, I can't help but notice the hubris of relying on technology when ecology has been here all along. Plants, fungi, and lichens were drawing carbon dioxide out of the air as early as 700 million years ago. Microbes have been quietly driving the Earth's carbon cycle, with little fanfare and a lot of humility.

We have been so focused on finding the elusive technological silver bullet that we turn a blind eye to transformations happening right where we stand, a climate solution rooted in soil that can be scaled to almost every acre of farm and ranchland. Maybe we can learn a lesson or two from the microbes and the ecological relationships they support. These tiny beings have been doing their thing since life on Earth began, shaping the conditions necessary for plants and all the organisms that depend on plants for survival. They will be here when we are long gone. On days when I struggle to imagine what the world will look like when we elevate a new generation of climate champions and fully harness the breadth of climate solutions, I turn to the humble microbe for inspiration.

Notes from a Climate Victory Garden

LOUISE MAHER-JOHNSON

Rebalance: Greenhouse gases (CO_2, N_2O, CH_4, H_2O vapor) with photosynthesis.
Recognize: Plants cool by evaporation, ground cover, shade, and precipitation.
Replant: Lawns with Victory Gardens, as in world war past.
Regenerate: Biodiverse farms with trees-flowers-herbs-pasture-animals.
Restore: Carbon out of air and back into soils, where it belongs.
Replace: Industrial monocultures with regenerative permacultures.
Revisit: Food production by many small farms, not a few megafarms.
Reject: Fossil fuel–based pesticides, plastics, and propaganda.
Rethink: Healthy ecosystems and economies for *all* life.
Relocalize: Slow food, slow lifestyles, and slow economies.
Rekindle: Simple and good, nature and nurture, feeling over thinking.
Refeel: Kinship with pivoting sunflowers and starry fireflies.
Revive: Wildness, woodlands, wetlands, wildlife, waterways.
Reestablish: Health of bees, butterflies, birds, bats, beetles.
Respect: Work of insects, both pollinating and recomposing life.
Remember: Everything is connected.
Everyone lives downstream and downwind.
Reimagine: Deep conservation, cooperation, and community.
Rebalance: Nature knows. Mimic her. Sense her. Be her.

Solutions at Sea

Emily Stengel

I'm here to tell you about the ocean. But I'll start with my roots, which are firmly on land.

When I was twelve, my first job was selling produce at a farmers' market in Pennsylvania's Amish Country. I was enchanted by the goods themselves—sweet strawberries freshly picked, spicy radishes that were perfect with the churned butter from the next stand over—but even more so by the people and the process. I worked for a three-generation family business whose owners took incredible pride in growing food for their community. For the Mecks, it wasn't a job but rather a vocation—a calling I deeply admired.

In my early twenties, my enchantment led to catering for a family-owned company in New York City, sourcing from the regional farm and food economy before "farm to table" was hip. I loved that work, coordinating with growers, chefs, servers, and clients, but felt pulled to go deeper into the workings of the food system.

I shifted my focus to agricultural research—academic work that took me around the country as I interviewed more than two hundred farmers in an effort to understand what made their farms successful. My eyes opened. People told me they were having trouble making ends meet because of struggles related to household issues, such as affording health insurance while working one of the deadliest jobs in America or finding appropriate childcare in rural communities. The farming population was aging out with no succession plan, and farmland was too expensive for young farmers to buy. Agriculture is often portrayed as an idyllic lifestyle; what I saw and heard painted a disheartening picture of inequity and persistent challenges for young, beginning, and women farmers in particular.

Beyond this perfect storm of socioeconomic challenges, I also started to hear from farmers about climate change and how it weighed on them.

Climate change imperils our ways of growing and our food system as a whole. Wildfires are scorching farmland from Australia to California. Floods are wreaking havoc on America's breadbasket. With agriculture using 90 percent of the world's freshwater resources and rising populations putting pressure on farmable land, water and land shortages will steadily drive up the cost of food production, making our current agricultural system increasingly untenable.

The challenges faced by our farming communities—compounded by the growing doom and gloom of climate change—felt insurmountable to me. The outlook was grim, and I struggled to find a way to contribute.

In the midst of this searching, a mutual friend introduced me to Bren Smith, a salty ex–commercial fisherman who had remade himself as an ocean farmer, raising shellfish and seaweeds off the coast of Connecticut. I didn't know anyone in that industry or even what it meant to "farm the ocean." Still, our friend thought I might want to help him teach his model to new farmers.

I read up on Bren before we met: A *New Yorker* profile described him as living in an Airstream and drinking water from a whiskey bottle. I wasn't sure I wanted to work with this guy, but I definitely wanted to meet him. In our first conversation, I was intrigued by his hopeful ambition, which went far beyond farming. We could mitigate climate change, restore ecosystems, and revive coastal economies through a simple but holistic partnership with the ocean.

I had never considered the ocean to be more than a place to swim, and I was being asked to consider ocean farming as a solution to the dual crises of access to affordable farmland and ecological sustainability that often blight earthbound agriculture and the food system it supports.

As Bren walked me through the regenerative ocean farming model, my doubt loosened and a paradigm shift was set in motion. I began thinking: Could farming at sea be a way to address the problems I was seeing on land? Could it offer opportunities for novice farmers, just starting out? Could it produce food without producing pollution?

In short, the answer is yes.

To grasp the benefits of ocean farming, it helps to start with the basics. Imagine a rectangular underwater garden. Hurricane-proof anchors on opposite edges are connected by horizontal ropes floating six feet below the surface. From these lines, kelp and other seaweeds grow vertically downward, next to scallops in hanging nets and mussels in mesh socks. Oysters sit below them in cages on the seafloor, with clams buried in the ocean's muddy bottom. It's an architecture of sustenance, nearly invisible above water.

These crops require zero inputs—no freshwater, no animal feed, no fertilizer, no pesticides. They're also restorative. Seaweeds such as kelp are often called the "sequoia of the sea" because, like sequoia trees, they are heroes of carbon sequestration. Oysters filter up to fifty gallons of water per day, removing nitrogen that seeps into our waters from land-based agricultural runoff, a major factor in aquatic dead zones.

The simple underwater rope scaffolding is cheap and easy to build, so this farming model is replicable and scalable. Once you secure a lease for a farm site, you can have a commercial-scale farm up and running within a year, with investment as low as $20,000. The demand for ocean-farmed crops continues to grow. Seaweeds can be used for food but also for fertilizers, animal feeds, and even bioplastics.

After learning, researching, reflecting, and considering the partners Bren had already engaged, I wanted in.

In 2016 I joined Bren to get our nonprofit, GreenWave, off the ground. We have spent the past four years building an organization that can support the growing interest in, and dire need for, this way of farming. To date, more than five thousand ocean-farmers-to-be from more than one hundred countries have expressed interest and want support.

Our direct work with the ocean farming community has illuminated what can make this field successful. Of course farmers need the basics: affordable seed, a lease and permit for their farm site, training and technical assistance, and connections to buyers. But we've learned that they flourish in clusters of several dozen farms that collaborate, learn together, and share land-based infrastructure, including farmer-owned hatcheries, cooperative processing hubs, and committed wholesale and institutional buyers.

Partner companies are building a market for ocean farmers' crops by developing climate-positive products such as food, fertilizers, livestock feed, and bioplastics. They are creatively incorporating local sources of seaweed into new and familiar items, from kelp jerky to seaweed pasta to drinking straws. There is also a growing market for seaweed fertilizer and compost, which create a virtuous nutrient loop, harnessing ocean carbon, nitrogen, phosphorus, and more and returning them to shore for sequestration in soil. Seaweed may also help address greenhouse gas emissions from livestock. One study * shows that supplementing livestock feed with a small amount of seaweed can reduce methane output by more than half in cattle.

Because regenerative ocean farming is in its infancy, we have a chance to do this right—to create a future where climate resilience and equity drive decision making in a new blue-green economy. This means focusing on *who* is farming, increasing opportunity for underrepresented communities including Indigenous peoples and women. The blue-green economy requires growing what is restorative and ecologically sustainable, not simply what consumers demand. It succeeds by collaborating and building circular business models that cut waste and keep valuable resources in use. It flourishes through philanthropic and legitimate impact investments of capital.

As the field grows, an interesting trend is emerging: Women are taking the helm. Men have historically dominated the marine space, but it's women who are surfacing as the captains of this new regenerative economy. They are farmers, hatchery technicians, scientists, and entrepreneurs. They are coming from diverse backgrounds of fishing, research, shellfish farming, science and technology, publishing, and marketing, to name just a few.

It makes sense. Our collaborative, holistic, and inclusive approach is distinctly feminine. We're working to renew coastal communities, to ally with public, private, and farming partners, and to ensure that the needs of both people and planet are considered. The synergistic design of our work isn't trying to solve just one problem. It aims to solve many problems at once—to hold all of these problems in tension with one another and figure out which levers to pull when. (I know this tension all too well as a woman and, more recently, as a mother juggling a new baby, my partner, and this work.)

As I scan the "oceanscape" of regenerative farming, I see women leading with resilience, initiative, and integrity. I'm proud to be one of them.

In the face of climate change, we have to act both fast and at scale. * Regenerative ocean farming can deliver on both. A recent World Bank report states that farming regenerative species in less than 5 percent of U.S. waters could produce protein equivalent to three trillion cheeseburgers, create more than fifty million new jobs, and absorb ten million tons of nitrogen and 135 million tons of carbon per year—all with no freshwater or chemical inputs. Today we're farming in just 0.004 percent of all coastal oceans. The opportunity is immense.

Working in partnership with our oceans, we can build an equitable blue-green economy that nourishes both people and planet. We can produce healthy food, replace plastics, and reduce the harms of land-based agriculture. We can boldly act on climate change, with women at the helm. We can wisely steward the solutions the sea is offering up.

Characteristics of Life

Camille T. Dungy

A fifth of animals without backbones could be at risk of extinction, say scientists.

—BBC Nature News

Ask me if I speak for the snail and I will tell you
I speak for the snail.
I speak of underneathedness
and the welcome of mosses,
of life that springs up,
little lives that pull back and wait for a moment.

I speak for the damselfly, water skeet, mollusk,
the caterpillar, the beetle, the spider, the ant.
I speak
from the time before spinelessness was frowned upon.

Ask me if I speak for the moon jelly. I will tell you
one thing today and another tomorrow
and I will be as consistent as anything alive
on this earth.

I move as the currents move, with the breezes.
What part of your nature drives you? You, in your cubicle
ought to understand me. I filter and filter and filter all day.

Ask me if I speak for the nautilus and I will be silent
as the nautilus shell on a shelf. I can be beautiful
and useless if that's all you know to ask of me.

Ask me what I know of longing and I will speak of distances
between meadows of night-blooming flowers.
I will speak
the impossible hope of the firefly.

You with the candle
burning and only one chair at your table must understand
such wordless desire.

To say it is mindless is missing the point.

Black Gold

Leah Penniman

Dijour Carter refused to get out of the van parked in the gravel driveway at Soul Fire Farm in Grafton, New York. The other teens in his program emerged, skeptical, but Dijour lingered in the van with his hood up, headphones on, eyes averted. There was no way he was going to get mud on his new Jordans and no way he would soil his hands with the dirty work of farming. I didn't blame him. Almost without exception, when I ask Black visitors to the farm what they first think of when they see the soil, they respond with "slavery" or "plantation." Our families fled the red clays of Georgia for good reason—the memories of chattel slavery, sharecropping, convict leasing, and lynching were bound up with our relationship to the Earth. For many of our ancestors, freedom from terror and separation from the soil were synonymous.

While the adult mentors in Dijour's summer program were fired up about this field trip to a Black-led farm focused on food justice, Dijour was not on board. I tried to convince him that although the land was the "scene of the crime," as Chris Bolden Newsome, farm manager at Sankofa Farm, put it, she was never the criminal. But Dijour was unconvinced. It was only when he saw the group departing on a tour that his fear of being left alone in a forest full of bears overcame his fear of dirt. He joined us, removing his Jordans to protect them from the damp earth and allowing, at last, the soil to make direct contact with the soles of his bare feet. Dijour, typically stoic and reserved, broke into tears during the closing circle at the end of that day. He explained that when he was very young, his grandmother had shown him how to garden and how to gently hold a handful of soil teeming with insects. She had died years ago, and he had forgotten these lessons. When he removed his shoes on the tour and let the mud reach his feet, the memory of her and of the land traveled from the earth, through his soles, and to his heart. He said that it felt like he was "finally home."

Our Sacred Ancestral Relationship with Soil

The truth is that for thousands of years Black people have had a sacred relationship with soil that far surpasses our 246 years of enslavement and 75 years of sharecropping in the United States. For many, this period of land-based terror has devastated that connection. We have confused the subjugation our ancestors experienced on land with the land herself, naming her the oppressor and running toward paved streets without looking back. We do not stoop, sweat, harvest, or even get dirty because we imagine that would revert us to bondage. Part of the work of healing our relationship with soil is unearthing and relearning the lessons of soil reverence from the past.

We can trace Black people's sacred relationship with soil back at least to the reign of Cleopatra in Egypt beginning in 51 B.C.E. Recognizing the earthworm's contributions to the fertility of Egyptian soil, Cleopatra declared the animal sacred and decreed that no one, not even a farmer, was allowed to harm or remove an earthworm for fear of offending the deity of fertility. Worms of the Nile River Valley are thought to have been a significant contributor to the extraordinary fertility of Egyptian soils. In West Africa, the depth of highly fertile anthropogenic soils serves as a "meter stick" for the age of communities. Over the past few centuries, women in Ghana and Liberia have combined several types of waste—including ash and char from cooking, bones from meal preparation, by-products from processing handmade soaps, and harvest chaff—to create African Dark Earths. This black gold has high concentrations of calcium and phosphorus, as well as 200 to 300 percent more organic carbon than soils typical to the region. Today community elders measure the age of their towns by the depth of the black soil, since every farmer in every generation participated in its creation.

When the colonial governments in northern Namibia and southern Angola attempted to force Ovambo farmers off their land, they offered what they said were equivalent plots with better-quality soil. The farmers refused to be displaced, countering that they had invested substantially in building their soils and doubted that the new areas would ever equal their existing farms in fertility. The Ovambo people knew that soil fertility was not an inherent quality but something that is nurtured over generations through mounding, ridging, and the application of manure, ashes, termite earth, cattle urine, and muck from wetlands.

This reverent connection between Black people and soil traveled with Black land stewards to the United States. In the early 1900s, George Washington Carver was a pioneer in regenerative farming and one of the first agricultural scientists in the United States to advocate for the use of leguminous cover crops, nutrient-rich mulching, and diversified horticulture. He wrote in *The American Monthly Review of Reviews* that the soil's "deficiency in nitrogen can be met almost wholly by the proper rotation of crops, keeping the legumes, or pod-bearing plants, growing upon the soil as much as possible." He advised farmers to dedicate every spare moment to raking leaves, gathering rich earth from the woods, piling up muck from swamps, and hauling it to the land. Carver believed that "unkindness to anything means an injustice done to that thing," a conviction that extended to both people and soil.

The Impacts of Our Estrangement from Soil

One of the projects of colonization, capitalism, and White supremacy has been to make us forget this sacred connection to soil. Only when that happened could we rationalize exploiting it for profit. As European settlers displaced Indigenous peoples across North America in the 1800s, they exposed vast expanses of land to the plow for the first time. It took only a few decades of intense tillage to drive around 50 percent of the original organic matter from the soil into the sky as carbon dioxide. The agricultural productivity of the Great Plains decreased 64 percent after just twenty-eight years of tillage by Europeans. The initial rise in atmospheric carbon dioxide levels was due to the oxidation of soil organic matter through plowing. *That means human-caused climate change started not just with the Industrial Revolution but with the exploitation of the soil.*

The planet's soils continue to be in trouble. Each year we lose around 25 million acres of cropland to soil erosion worldwide. The loss is ten to forty times faster than the rate of soil formation, driving carbon into the atmosphere and putting global food security at risk.

Soil degradation alone may decrease food production by 30 percent over the next fifty years.

When soils suffer the most egregious abuse, they can no longer even provide stable ground beneath our feet. In late 2018, wildfires blazed through parts of California, burning up the soil organic matter and ravaging the vegetation that held the hillsides in place. Heavy rain followed the Camp Fire, and the destabilized mud and boulders flowed downhill, leaving at least eighty-five dead and nearly nineteen thousand structures damaged or destroyed in its wake. Both the wildfires and the erratic rainfall can be linked to anthropogenic climate change and our voracious appetite for fossil fuels. Coupled with that, the process of extracting those fossil fuels from the Earth through coal mining and fracking further destabilizes the soil, resulting in sinkholes like the one in Chester County, Pennsylvania, connected to the Mariner East gas pipeline.

When the soil suffers, it's not just our food supply and our climate that are at risk. The further the population gets from its connection to Earth, the more likely we are to ignore and exploit those who work the soil. As Wendell Berry wrote in *The Hidden Wound:*

> The white man, preoccupied with the abstractions of the economic exploitation and ownership of the land, necessarily has lived on the country as a destructive force, an ecological catastrophe, because he assigned the hand labor, and in that the possibility of intimate knowledge of the land, to a people he considered racially inferior; in thus debasing labor, he destroyed the possibility of a meaningful contact with the earth. He was literally blinded by his presuppositions and prejudices. Because he did not know the land, it was inevitable that he would squander its natural bounty, deplete its richness, corrupt and pollute it, or destroy it altogether. The history of the white man's use of the earth in America is a scandal.

* In the United States today, nearly 85 percent of the people who work the land are Hispanic or Latinx and do not enjoy the same labor protections under the law as other American workers in other sectors. Pesticide exposure, wage theft, uncompensated overtime, child labor,

lack of collective bargaining, and sexual abuse are all-too-common experiences of farmworkers today. Record heat waves, attributable to climate change, have caused injury and death to farmworkers.

Society's abuse of soil and atmosphere has led to dire consequences for communities of color across the globe, who are disproportionately harmed by climate change. Devastating hurricanes have become regular annual visitors in the Caribbean islands and coastal areas of the United States. Several Alaskan Native communities struggle to hunt and fish in their traditional ways because rising temperatures are ravaging ecosystems and wildlife. And sub-Saharan Africa is among the regions projected to experience the harshest impacts of climate change. "If you're not affected by climate change *today*, that itself is a privilege," climate activist Andrea Manning says.

Black Farmers Heal the Soil and the Climate

But the same communities on the front lines of climate impacts are also on the front lines of climate solutions. A new generation of Black farmers is using heritage farming practices to undo some of the damage first brought on by the intense tillage of early European settlers. Their practices drove around half of the organic matter from the soil into the sky as carbon dioxide. Agriculture continues to have a profound impact on the climate; along with forestry, deforestation, and other land use, it contributes roughly 24 percent of global greenhouse gas emissions.

Now Black farmers are using heritage practices to reduce emissions and to capture excess carbon from the air and trap it in the soil. Our ancestral strategies are bolstered by Western science and listed among the most substantive solutions to global warming, per Project Drawdown's analysis.

One practice, silvopasture, is an Indigenous system that integrates nut and fruit trees, forage, and grasses to feed grazing livestock. Another, regenerative agriculture, involves minimal soil disturbance, organic production, compost application, the use of cover crops, and crop rotation. Both systems harness plants to capture greenhouse gases. Plants are nature's alchemists, transforming atmospheric carbon dioxide into sugar and trapping it on the land where it belongs.

Here are examples of how three womxn-led farms are putting these practices to work.

High Hog Farm, Grayson, Georgia

Not everyone in the Black farming community is as excited about fiber as Keisha Cameron. Given the prominent role of the cotton industry in the enslavement of African Americans, many farmers eschew cultivation of textiles. "We are largely absent from the industry on every scale," she explains. "Yet these agrarian art ways and life ways are part of our heritage."

At High Hog Farm, Cameron and her family are working to establish tree guilds, a system where fruit trees are surrounded by a variety of crops, in this case indigo, cotton, and flax. They also raise heritage breeds of sheep, goats, rabbits, horses, chickens, and worms in an integrated silvopasture system and sell fiber and meat. One of her favorite varieties is American Chinchillas, rabbits that consume a wider diversity of forage than goats and fertilize the pasture with their manure.

Their goal is a "closed loop" where all the fertility the farm needs is created in place. They pack a lot of enterprises into a small space. "We have five acres," she says playfully. "Just enough to be dangerous."

In his book *The Carbon Farming Solution*, Eric Toensmeier writes that silvopasture systems can absorb more than two metric tons of carbon per acre every year, stored in above- and below-ground biomass of grasses, shrubs, and trees. Emissions released by grazing animals can be completely offset by the carbon sequestered in well-managed pastures.

In addition to healing the climate, silvopasture is a joyful practice. "I get to play with sheep and bunnies. What could be better?" Cameron poses.

Soul Fire Farm, Grafton, New York

As Larisa Jacobson, codirector of Soul Fire Farm, explains, "Our duty as earthkeepers is to call the exiled carbon back into the land and to bring the soil life home."

When we first arrived on the land in 2006, the soil was tired. Nearly all of the topsoil had washed down the hill or been burned up into the sky. Our shovel tips met hardpan, gray clay, and the soil tests revealed a meager 3 to 4 percent organic carbon. Like so many Black farmers who have been historically dispossessed of land, we were fortunate to be able to afford even a few acres of marginal soil. We would make do.

The egg-laying chickens were the first contributors to the soil healing, pecking and scratching while depositing their rich manure over the land. We then added layers of compost, forest leaves, hay, and brown paper to form mounds for planting. As George Washington Carver advocated, we diligently planted leguminous cover crops wherever bare soil could be found, transmuting atmospheric carbon and nitrogen into solid form. Under Larisa's leadership, we minimized tractor tillage and used innovative but low-tech methods like placing tarps over empty beds to discourage weeds. As a result, the organic matter on the farm is now at precolonial levels of 10 to 12 percent, the native biodiversity is returning, and the soils are black, rich, and yielding.

My nickname on the farm is "Perennial Papa," because of my love affair with plants whose life cycles extend across many seasons. Unlike annuals, which are born, mature, and die in just one year, perennial plants know how to rest through the long winters and resume their growth in the spring thaw. Apples, blueberries, bee balm, mint, elderberry, peaches, and strawberries are among the myriad perennials I planted on the farm with the goal of trapping even more carbon beneath the soil. Nature abhors a monoculture, which is a large expanse of one single crop. Instead, nature plants trees together with shrubs, forbs, and grasses and has animals grazing through the system. We mimic nature's technology by creating our own agroforestry system, which may be the most climate-healthy type of farming possible.

Soul Fire Farm is a training center where thousands of aspiring Black and Brown farmers attend courses in Afro-Indigenous regenerative agriculture. Our goal is to catalyze and support a new generation of climate-savvy farmers committed to justice and healing.

Our ancestral grandmothers braided seeds and hope into their hair before being forced to board transatlantic slave ships, believing against odds in a future on soil. If they did not give up on us, their descendants, in those trying times, who are we to abandon faith? So we plant our seeds as well.

Fresh Future Farm, North Charleston, South Carolina

When Germaine Jenkins first moved to Charleston, she relied on SNAP and food pantries to feed her children. "I did not like that we couldn't choose what we wanted to eat," she says, "and there were few healthy options. I was sick of standing in line and decided to grow my own stuff."

Jenkins learned how to cultivate her own food through a master gardening course, a certificate program at Growing Power, and online videos. She promptly started growing food in her yard and teaching her food-insecure clients to do the same through her work at the Lowcountry Food Bank. In 2014 Jenkins won an innovation competition and earned seed money to create a community farm.

Today, Fresh Future Farm grows on 0.8 acres in the Chicora neighborhood and runs a full-service grocery store right on site. "We are living under food apartheid," explains Jenkins. "So all of the food is distributed right here in the neighborhood on a sliding-scale pay system."

Jenkins relies on what she calls "ancestral muscle memory" to guide her regenerative farming practices. Fresh Future Farm integrates perennial crops such as banana, oregano, satsuma, and loquat together with annuals like collards and peanuts. The farm produces copious amounts of compost using waste products like crab shell, and the farmers apply cardboard and wood chips in a thick layer of mulch. "We repurpose everything—old Christmas trees as trellises and branches as breathable cloche for frost-sensitive crops." Jenkins explains. They even have grapes growing up the fence of the chicken yard so that the "chickens fertilize their own shade."

Jenkins's farming methods have been so successful at increasing the organic matter in the soil that the farm no longer needs irrigation. It is also less vulnerable to flooding. "Two winters ago, we had four inches of snow. Our soil absorbed all of it," Jenkins says.

Toensmeier writes that for every 1 percent increase in soil organic matter, we sequester roughly 8.5 metric tons of atmospheric carbon per acre. If all of us were to farm like Jenkins, Jacobson, and Cameron, we could potentially put 300 billion metric tons of carbon dioxide back in the soil where it belongs. To keep warming to 1.5 degrees Celsius, we need both dramatic emissions reductions and widespread carbon sequestration; shifting our agricultural practices could comprise 15 percent of the puzzle, according to Project Drawdown.

Soil and Healing

The soil stewards of generations past recognized that healthy soil is not only imperative for our food and climate security—it is also foundational for our cultural and emotional well-being. My teachers, the Queen Mothers of Odumase Krobo, Ghana, admonished, "How can it be that you Americans put a seed in the ground, and you do not pray, sing, dance, or pour libations and you expect the Earth to feed you? The Earth is a relative, not a commodity. That is why you are all sick!"

Even Western science agrees that part of our sickness is connected to estrangement from the soil, positing that exposure to the microbiome of a healthy soil offers benefits to mental health that rival antidepressants. After mice were treated with *Mycobacterium vaccae*, a friendly soil bacteria, their brains produced more of the mood-regulating hormone serotonin. Some scientists are now advocating that we play in the dirt to care for our psychological health.

We see the benefits of soil anecdotally on our farm with the youth and adult participants who come to learn Afro-Indigenous soil-regeneration methods. While the curriculum focuses on such nerdy details as the correlation between earthworm count and soil organic matter, participants often reflect that the main thing they gain from their time with the dirt is "healing" and the strength to leave behind addictions, toxic relationships, poor diets, and demeaning work environments.

Our ancestors teach us that it's not just soil bacteria that contribute to this healing process. Part of African cosmology is that the spirits of our ancestors persist in the Earth and transmit messages of encouragement and guidance to us through contact with the soil. Further, we believe the Earth herself is a living, conscious spirit imparting wisdom. When we regard a handful of woodland soil, rich in the mycelium that transmits sugars and messages between trees, we are made privy to the inner world of the forest superorganism and its secrets of sharing and interdependence.

In healing our relationship with soil, we heal the climate, and we heal ourselves.

Like Dijour, we are welcomed home to a profound web of belonging that extends beyond the boundaries of self and species. One student on our farm reflected, "I leave this experience feeling grounded like a tree in a land and country that I previously did not feel welcomed in. Connection with soil was the awakening of my sovereignty."

Ode to Dirt

Sharon Olds

Dear dirt, I am sorry I slighted you,
I thought that you were only the background
for the leading characters—the plants
and animals and human animals.
It's as if I had loved only the stars
and not the sky which gave them space
in which to shine. Subtle, various,
sensitive, you are the skin of our terrain,
you're our democracy. When I understood
I had never honored you as a living
equal, I was ashamed of myself,
as if I had not recognized
a character who looked so different from me,
but now I can see us all, made of the
same basic materials—
cousins of that first exploding from nothing—
in our intricate equation together. O dirt,
help us find ways to serve your life,
you who have brought us forth, and fed us,
and who at the end will take us in
and rotate with us, and wobble, and orbit.

Water Is a Verb

Judith D. Schwartz

Katherine and Markus Ottmers live and work in the dry steppes of far-western Texas. Since rains can be few and far between, they've designed the main building at Casa de Mañana to collect both rainwater and condensation. But they had no idea how much water they harvested solely from dew until the winter morning when the valve burst on one of the water tanks.

Markus was doing ironwork outside when he noticed water gushing from the tank. "Hey, Brad!" he called to his coworker. "Go see how much water is in there. It can't be full. We haven't had any rain in four months." Not only were they rainless; they'd been providing for a herd of fifty goats, and between six and eight people were regularly taking light showers there. Conscious of every bit of water used, he knew it was in the neighborhood of fifty to seventy gallons a day.

Brad checked, and the tank was completely full. The Ottmerses continued to monitor the tank, rising at four thirty the next morning to check its status. Sunrise was yet hours away and stars saturated the sky. This sparsely populated area puts on a night sky as clear as you can find stateside—not just the bright starry dots but the smudgy sweep of distant galaxies in between. And here, at the edge of Big Bend, in the midst of a multiyear drought, a bleary-eyed Markus discovered that water was streaming into the tank at a rate that amounted to about sixty gallons a day, enough to cover nearly all of their water needs. Says Markus: "I was wondering who the water fairy was."

The Ottmerses understood that there is always moisture in the air, silently sailing above us. According to Brazilian scientist Antonio * Donato Nobre, the amount of water in the "aerial river" above the Amazon rainforest exceeds the water flowing in the Amazon River itself. Moisture hovers above even the driest landscapes. In planning their "rain barn," Katherine and Markus determined that by amplifying the temperature differential—between cool evening breezes and the sun-warmed metal roof—they could "capture" this moisture and have plenty of water to use, even as their neighbors fretted over the lack of it. They were inspired by the Namib Desert beetle, which drinks by catching droplets of water from the fog. This clever creature stands on its forelegs and the water rolls down its belly and into its mouth, quenching its thirst. As Rachel Carson wrote in an essay called "Clouds," "Up there is another ocean."

The point is, if you know how water works, you don't need a water fairy.

The unseen moisture that flows through the sky is important not only as a water source. It also conveys heat. In fact, the transfer of thermal energy via the phase changes of water is the primary means by which our planet manages temperature. Water is in constant flux, moving through the air and between forms, shape-shifting from gas to liquid to solid and back again.

This gives us a glimpse into the dynamism of air, that which we breathe and move through, a kind of mirror ocean riding ambient tides. Day and night, wafts of vapor wheel around the atmosphere, transforming conditions on the ground despite its weightlessness and invisibility. This moisture is, inevitably, an essential part of what creates and modifies our climate. Yet this humidity is not simply recycling moisture in a tidy, predictable way. It is diffuse and capricious, on the move, and determined by other equally mysterious factors, including the presence and vigor of plants.

When we make the connection between water and climate, the discussion tends to go in one direction: the many ways that climate change will bear on water. For example, since warmer air holds more water, with higher temperatures we are subject to heavier rains and more intense storms, plus the advent of sea level rise and the increased frequency and severity of droughts. What's been missing from the conversation is *the impact of water on climate*. While carbon dioxide traps heat, water vapor transports it, alternately holding and releasing thermal energy as it circulates. From that standpoint, we can think of water as the most significant greenhouse gas.

It is time to bring water into our climate strategies. Because, as it turns out, water is a climate ally: There is much we can do to influence how water moves across landscapes and through the atmosphere and thereby helps regulate heat. As the Ottmerses' experience demonstrates, we have plenty of opportunities to work with the water cycle instead of against or outside it.

I'd met Katherine in Albuquerque at the Quivira Coalition conference, an annual gathering devoted to enhancing western landscapes, and was immediately intrigued by the way she talked about water. She told me about "guzzlers," instances like that day when condensation caused the water tank to overflow. She said that in the desert, "we don't have a lot of rain, but we have 'moisture events.'" She told me about "nutrient-dense fogs" with morning mists so thick "you can't see the truck in the driveway" and how such events kindle something in the vegetation: "The plants are a lot happier with that little bit of moisture." She explained how she situates her plants so that they get morning shade and hold the dew longer into the day. I was charmed by her evocative use of language and unusual way of looking at things. "We're so sensitive to water that we can tell the little shifts in the moisture in the air," she said. "We're like the desert plants that way."

We generally think of water as a noun, as something bounded by place: a lake, river, or reservoir; that which comes out of a tap. Katherine helped me understand that water is also a *verb*: expanding in volume or retrenching; changing state in an ongoing dialogue with land and sun. This is not just to fuss over language. Rather, I believe that understanding how water works is essential to address our many water challenges. This is so whether we're contending with scarcity, in the case of drought, or too much, as in floods. And because the workings of water intersect with climate, biodiversity, and food security, zeroing in on water processes can help us grapple with our major global problems.

Let's take a quick look at how water behaves.

First, *infiltration*. In a healthy landscape, rain is held in the ground, where it supports plant and microbial life or slowly filters into groundwater stores. Our water "infrastructure" here is soil, and the richer the better: Every 1 percent increase in soil organic matter (which is mostly carbon) represents an additional twenty thousand gallons of water per acre held in the ground. With all this water soaking in, floods become a lot less likely, and less irrigation is needed since the soil stays moist between rains.

What we perceive as a lack-of-water problem is often an inability to keep water in the ground—a symptom of soil that's lost its carbon. (A significant percentage of atmospheric carbon dioxide is the result of soil-carbon depletion. As a colleague puts it, "Nature wants her carbon back.") My friend Precious Phiri, based in Zimbabwe, where she is the Africa coordinator for Regeneration International, says, "There are places where you will be in drought no matter how much rain you get." Simple approaches to building soil carbon, such as managed animal impact, can bolster land's water-holding capacity, which in turn boosts food security. In Zimbabwe I visited communities where Precious has trained villagers in holistic cattle grazing. Increased water infiltration in animal-treated fields has meant they can grow food for seven months of the year rather than merely two—the difference between being self-sufficient and relying on international food aid.

Going back to that water streaming into the Ottmerses' water tank, we might wonder: Where does all that water in the air come from? The source is a combination of evaporation from bodies of water and the land surface but mostly, between 80 and 90 percent, from *transpiration*. This is the upward movement of water through plants. You can think of it as the plant "sweating": The stomata on the leaves (or blades, if grass) open to retain or release moisture to cool the plant. What's important is that this is a *cooling* mechanism: transforming solar energy into latent heat—suspended in vapor—as opposed to sensible heat, or heat you can feel, like a hot sidewalk.

Consider a tree, say, a good-sized tree in full leaf, worthy of sitting under on a summer afternoon. On a sunny day, when it's basking in light, our tree will transpire more than twenty-six gallons of water. (Nobre often refers to trees as "geysers.") This dissipates solar energy. According to Czech plant physiologist Jan Pokorny, the transpiration action of one tree on a sunny day represents three times the cooling capacity of an air-conditioning system in a five-star hotel room. When we enjoy the gentle cool of a forest, it is thanks not just to shade but also to the botanical work of all those trees transpiring, what Pokorny calls "the world's most perfect air conditioner." *

We regard plants mainly as *recipients* of water—but they are also key determinants of *where water goes* and *what it does*. In other words, vegetation, particularly trees, helps drive weather and climate. Australian Peter Andrews, who developed Natural Sequence Farming, a land-management approach to make the best use of water on that dry continent, makes a point of this. In his book *Back from the Brink: How Australia's Landscape Can Be Saved*, he writes: "Every plant is a solar-powered factory producing the organic material on which all life depends. Every plant is also a pump, which is constantly raising water from the ground to keep the factory running."

Plants' capacity to regulate temperature—cooling during the day and closing stomata to conserve heat at night—has not figured in climate discussions and policies. And yet, Andrews says, "each day the planet takes in a certain number of units of heat, which it somehow has to manage. In the past, billions of plants helped to manage the heat in situ. . . . Around a quarter of the planet has now been stripped almost entirely of vegetation. In other words, one quarter of the planet has been stripped of its ability to moderate temperature." However, he adds, "The fact that all the major problems of our landscape have a common cause, a lack of vegetation, means that they also have a common solution."

Now, finally, the water action the Ottmerses promoted at their rain barn: *condensation*. Most simply, condensation is the process of water in its gaseous form turning to liquid, what scientists call a "phase transition." Condensation is all around us. It's when you pick up your cool drink on a summer day and water droplets form on the glass. It's when you step out onto grass early in the morning and your feet get wet. The condensation part happens when warm and cold collide, as when the nighttime air hovers over a pond whose surface has been warmed by the sun and a bank of fog materializes, or when the evening breeze circulates between the sunbaked upper roof and the shaded plane below the Ottmerses' barn.

Condensation can be seen as the reverse of transpiration, its meteorological mirror: Transpiration absorbs latent heat; condensation releases this withheld heat. This doesn't necessarily mean a temperature increase. Rather, the conferred heat contributes to the amassing and ascent of clouds. When water shifts from liquid to gas, the molecules spread out. When water vapor becomes a liquid, the molecules retract and move closer together.

This process is at work every time it rains or snows. Here, too, plants play a pivotal role. Douglas Sheil, a tropical forest ecologist now based in Norway, wrote an article entitled "How Plants Water Our Planet" that begins: "Most life on land depends on water from rain, but much of the rain on land may also depend on life." Why is this so? For one thing, the source of rain is primarily water transpired from plants. But also, plants help bring the rain down.

While rain is a product of condensation, water can't do it alone. To produce a raindrop, water molecules need a surface to condense upon. Enter "condensation nuclei": the flecks of particulate matter around which moisture coalesces, mainly ice crystals, salts, pollen, and scented compounds produced by plants. Nobre notes that in the rainforest, tree leaves emit volatile organic compounds that trigger rain. He calls these "scents of the forest" or, inspired by the animated films his two daughters favor, "pixie dust."

These water processes are intertwined: Water rises to form vapor and returns as rain, replenishing the area from which it derived. Ideally, the moisture is used and reused before it flows into streams, into rivers, and ultimately back to the ocean. Only when water infiltrates the soil can plants grow and initiate the circuit that will ultimately bring the rain. The plants also send carbon into the soil, feeding microbial life and acting as a sponge, able to hold on to water and sustain plant and animal life.

When these cycles are broken, conditions deteriorate and the landscape becomes susceptible to drought, flooding, and wildfires. It has been unnerving to grasp the vulnerability of many ecosystems around the globe and then watch in real time as they become yet further damaged. The deforestation in Brazil is extremely disturbing, particularly once you consider the role of trees in cycling water and moderating heat in the region and beyond. The severe flooding in Midwest farm areas is at once a result of degraded soil and a contributor to further topsoil loss. Similarly, a failure to understand how to manage Australia's landscape—knowledge long held by largely disenfranchised Aboriginal communities—contributed to the recent devastating wildfires.

It is extremely painful to acknowledge the damage that's been done to the environment. At this difficult moment it is essential to embrace our humility, our place in the cosmos, and to accept what we cannot control. But it is also important to acknowledge where we *do* have leverage. One place where we have tremendous agency is with the water cycle, which, as we've seen, is largely driven by vegetation and the condition of soil. I think it important to call attention to the fact that we have so far neglected the role of functioning ecological systems in climate regulation. For it is such healthy environments that maintain the water cycle and therefore manage heat.

The beautiful thing is that there is nothing we have to "engineer." Nature does this for us. Basically, what we need is life: life to transport water and, by extension, to regulate heat; life to seed the rain; life to slow down moving water so that it has a chance to infiltrate. Creatures from earthworms and dung beetles to beavers and prairie dogs help create meanders so that water lingers in the landscape and soaks in. Microbes provide nutrients to plants so they enrich and stabilize the soil. When appropriately managed, herbivores like cattle and sheep help build carbon, the way the buffalo created our country's once-fertile prairies.

We can all find ways to help living things thrive, whether through joining or starting a Climate Victory Garden, reserving a section of lawn for pollinator-attracting plants, or supporting farmers who grow crops without pesticides. There is joy in acting upon our love for the Earth. Plus, improvements on every scale matter. Think of how much water is produced and heat dispersed by that single leafy tree.

In Texas, Katherine Ottmers says working with water in the landscape is rewarding—and creative. "I like to think of Casa de Mañana as one big art project. Our canvas is degraded land," she says. Not that she is naive about the state of the world. Rather than ruing the fact that humans are changing the Earth, she believes we need to embrace the responsibility this entails. "We can be the beavers on the landscape, the keystone species." She calls their land-restoring project "oasification," bringing water and life to a parched and wounded landscape. The sign at the rain barn's entrance reads "Bloom Where You Are Planted."

Nature wants to heal herself and will do so if given a chance or—better—a nudge.

The Seed Underground

Janisse Ray

Agriculture has created in us a story-based, community-reliant, land-loving people. It has given us a head start on what I call the Age of Bells, the time when bells—cowbells, dinnerbells, bells of flowers—will again be ringing across the hills and plains. We are coming to the new age of agriculture better prepared: knowledgeable about growing, able to do with less, happy in our communities, firm in gender and racial equality, healthier. I believe that the organic and local-food movement is leading the way to re-creating cultures that are vibrant and vital. What we are witnessing in agriculture is no less than a revolution.

It also means we are on an edge—lots of edges, in fact. When I think of the edge, I think first of a literal one, the fencerow, which modern chemical agriculture has been destroying. This is the place where birds pooped out wild cherry seeds and wild cherry trees grew; and the place where, tired from the row, workers sat in the shade and told stories. It's where a lone farmer watched a mockingbird sing.

We occupy an edge between forest and field, the most exciting place in the world to me. We are on many edges: balancing the needs of the wild with the need to nourish people, balancing urban life with the need to eat, balancing concerns about human health with the need for productivity, weighing input against output, and making decisions based on both ecology and economy.

There is also a psychological edge we're all living on. We know that we're living in a world that is being devastated but also one replete with the beauty and power of life. We live on the boundary of deciding to make positive contributions although we know we are complicit in the destruction. We skate between apathy, because the truth of what's happening is painful to think about, versus action, any kind of action; and we skitter between the paralysis caused by grief and fear versus action. Every decision we have to make, whether it's a life-sustaining or a life-destroying one, is an edge. Our very psyches are on the edge, between dropping out and dropping in, between selling out and fighting back. Every single one of us.

The verge is a dangerous and frightening place. It's important to know that one is not alone on it. The edge holds a tremendous amount of ecological and cultural as well as intellectual power. I believe that we have to get comfortable with it.

How shall we live? As if we believe in the future. As if every one of us is a seed, which as you know is a sacred thing. In my wildest dreams the seeds of every species are speaking to me, calling out: *In all the bare spots on earth plant us and let us grow. On all the edges, plant seeds.*

8. RISE

CAN YOU IMAGINE THE COMMUNITY THAT WILL HEAL THE CLIMATE CRISIS?

IT WILL NOT BE JUST YOU

IT WILL NOT BE A TECHNOLOGICAL SALVATION

IT WILL BE ALL OF US

MADELEINE JUBILEE SAITO

A Letter to Adults

Alexandria Villaseñor

Dear Grown-ups,

Now is the time to be a climate activist.

I am fifteen years old and spend a lot of my time on conference calls, sending emails, speaking publicly, and going to protests. Those are probably different memories than you were making at my age, but we youth know we need to make our voices heard now—because our generation will feel climate impacts the most. Scientists tell us that we have to reduce global greenhouse gas emissions by about half *this decade* to avoid irreversible, catastrophic effects of climate change.

The climate crisis is the largest generational inequality there is. Generations that came before mine had greater access to nature and natural resources than young people today will ever have. Rather than having abundant clean water, we face a global water crisis made worse by drought. Climate breakdown threatens global food supplies because storms, floods, and droughts make it harder to grow food. Globally, an estimated nine million people die every year from air pollution. The planet that young people are receiving is one stricken with disaster, and it's unfair.

But we are not going to sit by and watch as our ecosystems and climate system collapse. Young people are becoming activists because of the damage happening to their communities and the environment. My activism sparked in November of 2018, when I was visiting extended family in Northern California, where I was born and raised. The state's deadliest wildfire in history broke out, the Camp Fire near the town of Paradise. Over two hundred square miles burned, and about fourteen thousand homes were destroyed. The air where I was visiting reached over 350 on the Air Quality Index, a level deemed hazardous, inflaming my asthma. I ended up becoming very sick. For my health, my family sent me back home to New York City.

I was so upset about what had happened in my home state. I wanted to know what had caused the fire. I researched and started to see the connection between climate change and California's wildfires. I learned that climate scientists could calculate what's called the climate change signal out of events like the Paradise Fire. That signal tells us that climate change is intensifying wildfires and other weather events. When I learned what was happening to our planet, I was unable to ignore it.

A few weeks later, I decided to make my voice heard by going on a school strike for climate on December 14. I have been on strike every Friday since then.

As you, hopefully, know, the climate strike movement was started by a teenager in Sweden. Greta Thunberg, who was then fifteen, started striking every Friday in front of the Swedish parliament to demand action on the climate crisis. A global movement quickly exploded, and schoolchildren are now striking every Friday for climate action. When Greta started striking, she symbolically encouraged young people all around the world to make their voices heard. She inspired me to use mine.

At the time of writing this, I can't vote (though maybe that should change), which means I can't choose who we put in positions of power, even though I will inherit the decisions they make. So I'm taking action and demanding change in other ways.

Young activists around the world are doing whatever it takes to make our voices heard on the climate crisis. Beyond protests and civil disobedience, we are lobbying political leaders for climate action. We are engaging our local governments, civic leaders, teachers, and school administrations. We are demanding climate action in our communities and climate education in our schools.

We are also taking legal action. In 2015, twenty-one youth plaintiffs from around the United States filed a landmark case, *Juliana v. U.S.* The case asserts that the U.S. government contributes to climate change and is violating young people's rights to life, liberty, and property. The plaintiffs and lawyers continue their uphill fight. In October 2019, fifteen children from across Canada sued their government for supporting the fossil fuel industry, arguing that it is jeopardizing their rights as Canadian citizens.

There's an international legal case now too, which I have the honor of being a petitioner for. On September 23, 2018, Greta Thunberg, fourteen other children, and I filed a complaint with the UN Committee on the Rights of the Child. *Children vs. Climate Crisis* intends to challenge current ideas of national sovereignty and bring awareness to the global nature of the climate emergency humanity faces. This groundbreaking legal complaint states that five countries—Argentina, Brazil, France, Germany, and Turkey—are violating the Convention on the Rights of the Child because of their inaction on the climate crisis. This convention is the most widely ratified treaty on the planet, and it states that children have an inherent right to life and a supportive environment in which to grow. We hope that the five respondent countries not only begin to lead by example with their national climate actions but also bring the entire planet together to work on solutions.

A question I get asked a lot is "Do you miss just being a regular teenager?" And the answer is yes. I miss doing theater, playing volleyball, and hanging out with my friends. But the climate crisis threatens every aspect of my future. So what other choice do I have? It is a moral obligation to fight for this planet. My fight for climate action is not going to end until our planet and all its people are safe.

Already, people have had to leave their homes because of floods, hurricanes, wildfires. Already people are enduring disease outbreaks made worse by our warming planet. Already people's food supplies are at risk because of droughts, floods, and pest invasions. We can no longer ignore the fact that climate change is causing so much destruction across our planet right now.

What truly breaks my heart is that we are simultaneously going through an ecological collapse. Extinction rates are tens to hundreds of times higher than they have been in the past ten million years, and the climate crisis will likely make them worse. One species I would be particularly sad to lose is the monarch butterfly. Growing up in Northern California, they were everywhere. It was always a highlight of my spring, seeing all of the monarch butterflies fluttering around and sipping flower nectar. I remember visiting a butterfly farm, and they would land on my extended finger; a couple landed on my head. It made me so happy seeing these ethereal flying insects.

In the winter of 2018–19, the western monarch butterfly population dropped below thirty thousand—a drop of 86 percent in a single year. Between habitat loss, pesticides, and possibly climate change, monarchs may soon be added to the endangered species list. They are not the only species facing this fate. If I end up having children, I want them to be able to experience the vast biodiversity we have today, but I am worried that they won't get to have these experiences.

Because we are invested in a viable future, youth are stepping forward as the "conscience and moral voice" on the climate crisis. That is why we continually put ourselves in important decision-making spaces all over the world. We are here to push corporate and government leaders, and *all* of you adults, to do the right thing. We are doing this not only because the science says so, not only because we're headed for ecological and planetary catastrophe, but because we are human.

After a year of weekly local strikes, helping to organize a big global strike, and participating in international legal actions, I've learned a lot about climate activism. Most important, I've learned that to make change happen quickly we need everyone to be an activist!

The biggest reason why people don't act: They don't know about or understand the climate crisis. My generation needs to know the unique risks and vulnerabilities we are facing; that must be a part of our education. Since schools generally aren't doing this, I started Earth Uprising, a nonprofit that focuses on peer-to-peer climate education. Until you grown-ups get it together to improve school curricula, young people can teach one another what's happening to our planet and how to help mitigate the climate crisis.

When adults come to talk to me at protests or conferences, they often tell me that before witnessing this youth movement, they didn't have any hope for the future, but they now believe our generation will be the ones to save the planet and humanity. Many of you say how sorry you are for placing this burden on our shoulders—that it shouldn't be our responsibility to right the wrongs of the generations who have come before us.

But the reality is it is too much work for one generation. Those of you who are retired and have more time on your hands, or with children you are no longer caring for, or those of you with additional resources—consider becoming a climate activist. Can you imagine how beautiful a movement led by children and grandparents would be?

To me, the responsibility to save and protect our planet for ourselves and future generations is *not* a burden. I think it's a blessing, and I believe we were chosen for this job because we are the ones who can make it happen. In this unique moment in history, we must come together to save our future. Through our activism, strikes, and fearless speaking out, the youth of the world are calling on adults to care for all the glorious life that exists here—to do it now, before it's too late.

The climate crisis is the largest challenge humans have ever faced. We young people are doing everything we can, so please join us. We need your help.

Welcome to the uprising!
ALEXANDRIA

Science tells us that it's not too late, but we have to pull hard, every day, together, to make a difference. . . . You don't have to know where we'll end up. You just have to know what path we're on.

—Prof. Kim Cobb

An Offering from the Bayou

Colette Pichon Battle

It was about two years after Hurricane Katrina that I first saw the Louisiana flood maps. These flood maps are used to show land loss in the past and the land loss predicted to come. On this particular day, at a community meeting, these maps were used to explain how a thirty-foot storm surge that accompanied Hurricane Katrina could flood communities like mine in South Louisiana and communities across the Mississippi and Alabama coasts. It turns out that the land we were losing, including our barrier islands and marshland, was our buffer from the sea. The time-lapse graphic showed massive land loss in South Louisiana and an encroaching sea. But more specifically, the graphic showed the disappearance of my community and many other communities in South Louisiana before the end of the century.

I was standing there with other members of South Louisiana's communities—Black, Native, poor. We thought we were just bound by two years of disaster recovery, but we found that we were now bound by the impossible task of ensuring that our communities would not be erased by sea level rise due to climate change. Friends, neighbors, family, my community: I had just assumed they would always be there. Land, trees, marsh, bayous: I had just assumed that they would be there, as they had been for thousands of years. I was wrong, and my life forever changed.

To understand what was happening to my community, I talked with other communities around the globe. I started in South Louisiana with the United Houma Nation, a state-recognized tribe. I talked to youth advocates in Shishmaref, Alaska, whose barrier island is rapidly melting and eroding into the sea. I talked to fisherwomen in coastal Vietnam, justice fighters in Fiji, new generations of leaders in the ancient cultures of the Torres Strait. Communities that had been on Earth for thousands of years were suffering the same fate, and we were all contemplating how we would survive the next fifty.

By the end of the next century, it's predicted that 200 million people will be displaced due to climate change, and in South Louisiana those who can afford to do so are already moving. They're moving because South Louisiana is losing land at one of the fastest rates on the planet. Disappearance is what my bayou community has in common with other coastal communities. Erasure is what communities around the globe are fighting as we get real about the impacts of climate change.

I've spent the last fifteen years advocating on behalf of communities that have been directly impacted by the climate crisis. These communities are fighting discrimination within climate disaster recovery while trying to balance mass displacement of people with an influx of others who see opportunity in starting anew. I hear people being called "refugees" when they leave or when they're displaced by climate disaster, even when they don't cross international borders. These misused terms, meant to identify the other, the victim, the person who is not supposed to be here, are barriers to economic recovery, to social integration, and to the healing required from the climate crisis and climate trauma. Words matter.

It also matters how we treat people who are crossing borders due to disasters. We should care about how people who are seeking refuge and safety are being treated, if for no other reason than that it might be you or someone you love who needs to exercise their human right to migrate in the near future.

We must start preparing for global migration today. It's a reality now. Our cities and our communities are not prepared. In fact, our economic system and our social systems are prepared only to make profit off disaster and people who migrate. Climate-disaster migration will cause rounds of climate gentrification, and it will also penalize the movement of people, usually through exploited labor and criminalization.

Climate gentrification that happens in anticipation of sea level rise is what we're seeing in places like Miami, where communities that were once kept from the waterfront are being priced out of the high ground. The land farthest away from the oceanfront vista is where poor and Black and immigrant populations were placed originally. Now, as resourced people escape the long-term coastal impacts of climate change, poor communities are being forced to relocate, away from the social and economic systems that they need to survive.

Climate gentrification also happens in the aftermath of disaster. When massive numbers of people leave a location for an indefinite amount of time, we see others come in and claim it. We also see climate gentrification happen when damaged homes are reconstructed as "green built" and warrant a higher price tag, generally out of the reach of Black and Brown and poor people who want to return home. The jump in rents or home prices can mean the difference between being able to practice your human right to return home as a community and being forced to resettle somewhere else—less expensive, but alone.

The climate crisis is a much larger conversation than reducing carbon dioxide emissions, and it is a very different conversation from just extreme weather. We're facing a shift in every aspect of our global reality. Climate migration is just one small part, but it's going to have ripple effects in both coastal cities and cities in the interior.

So what do we do? I have a few ideas.

We must reframe our understanding of the problem. Climate change is not the problem. Climate change is the most horrible symptom of an economic system that has been built for a few to extract every precious ounce of value out of this planet and its people, from our natural resources to the fruits of our human labor. This system has created this crisis.

We must have the courage to admit we've taken too much. We cannot close our eyes to the fact that the entire world is paying a price for the privilege and comfort of just a few people on the planet. It's time for us to make society-wide changes to a system that incentivizes consumption to the point of global imbalance. Our social, political, and economic systems of extraction must be transformed into systems that regenerate the Earth and advance human liberty—globally. It is arrogance to think that technology will save us. It is ego to think that we can continue this unjust and extractive approach to living on this planet and survive.

To survive this next phase of our human existence, we will need to restructure our social and economic systems to develop our collective resilience. The social restructuring must be toward restoration and repair of the Earth and the communities that have been extracted from, criminalized, and targeted for generations. These are the front lines. This is where we start.

We must establish a new social attitude to see migration as a public benefit—a nature-wide necessity for our global survival—not as a threat to our individual privilege. Collective resilience means developing cities that can receive people and provide housing, food, water, healthcare, and freedom from overpolicing for everyone, no matter who they are, no matter where they're from.

What would it mean if we started to plan for climate migration now? Sprawling or declining cities could see this as an opportunity to rebuild a social infrastructure rooted in justice and fairness and a physical infrastructure that works to restore and preserve our ecology. We could actually put money into public hospitals to both help them rebuild in a more eco-friendly way and prepare for what is to come through climate migration, including the trauma that comes with loss and relocation. We must invest more in justice, but it cannot be for temporary gain. It cannot be to help budget shortfalls. It has to be for long-term change.

It's already possible, y'all.

After Hurricane Katrina, universities and high schools around the United States took in students to help them finish the semester or the academic year without missing a beat. Those students are now productive assets in our communities, and this is what our schools, our businesses, and our institutions need to get ready for now.

So as we reframe the problem in a more truthful way and we restructure our social systems in a more just way, all that will be left is for us to re-indigenize ourselves and to conjure a power of the most ancient kind.

This necessarily means that we must learn to follow—not tokenize, not exotify, not dismiss—the leadership and the traditional knowledge of a particular local place. It means that we must commit to standards of ecological equity and climate justice and human rights as a base standard, a fundamental starting point, for where our new society is to go.

All of this requires us to recognize a power greater than ourselves and a life longer than the ones we will live. It requires us to believe in the things that we are privileged enough not to have to see.

We must honor the rights of nature. We must advance human rights for all. We must transform from a disposable, single-use, individual society into one that sees our collective, long-term humanity, or else we will not make it. We must see that even the best of us are entangled in an unjust system, and we must acknowledge that *your* survival requires us to figure out how to reach a shared liberation together.

The good news is that we come from powerful people. We come from those who have, in one way or another, survived. This is reason enough to fight. And take it from your South Louisiana friend: Those hardest fights are the ones to celebrate. Let's choose to make this next phase of our planetary existence beautiful, and while we're at it, let's make it just and fair for everyone.

We can do this, y'all. We can do this, because we must. We must, or else we lose our planet and we lose ourselves. The work starts here. The work starts together. *Merci d'avoir reçu ce que j'offre.* Thank you for receiving my offering.

Calling All Grand Mothers

ALICE WALKER

We have to live
differently

or we
will die
in the same

old ways.

Therefore
I call on all Grand Mothers
everywhere
on the planet
to rise
and take your place
in the leadership
of the world

Come out
of the kitchen
out of the
fields
out of the
beauty parlors
out of the
television

Step forward
& assume
the role
for which

you were
created:
To lead humanity
to health, happiness
& sanity.

I call on
all the
Grand Mothers
of Earth
& every person
who possesses
the Grand Mother
spirit
of respect for
life
&
protection of
the young
to rise
& lead.

The life of
our species
depends
on it.

& I call on all men
of Earth
to gracefully

and
gratefully

stand aside
& let them
(let us)
do so.

A Field Guide for Transformation

LEAH CARDAMORE STOKES

> I live my life in widening circles
> that reach out across the world.
>
> —RAINER MARIA RILKE

I was more than a decade into being an environmental activist when I saw it: energy. It was everywhere. In the fuel for our cars, in the fertilizer for our food, in the plastics at the grocery store, in the electricity lighting our homes. We were relying on fossil fuels for everything. Once I learned to see them, I could find fossil fuels wherever I looked. And it wasn't hard to see their harms either: coal, oil, and gas make people sick, pump pollution into our air and water, and push up carbon concentrations in the atmosphere, driving dangerous weather. It took me a long time to understand that government policy has created this climate crisis. Our politicians subsidize fossil fuels while blocking clean energy. We need to turn this around. If we could just stop digging up fossil fuels, I thought, we might stand a chance.

But this realization did not come easily. My first environmental actions were not about changing our energy policy. Instead, they were small and immediate. After discovering clear-cutting in the Amazon, in fifth grade, I made a T-shirt that featured cattle talking in the rainforest. They said: "Do not eat me." When my friends and I found out that the tiny milk boxes we got at lunch could not be recycled at school, we stopped playing outside. Instead, we spent our lunch hour meticulously cutting up the cartons so we could lay them flat and bring them home to recycle them. We did this for weeks.

In my teens, my actions got bigger and I started to stretch beyond myself. I wrote to the owners of the local grocery store to ask them to stop selling Chilean sea bass—a vulnerable species. I got my family to stop using disposable plates at family gatherings. I would flip the bird at Hummers and other gas-guzzling trucks in a kind of public protest. But when my high school geography teacher introduced the idea of climate change, he told us it probably wasn't real. It took me years to figure out that fossil fuels are at the center of climate change.

When I went to university, I had my first real chance to learn about the climate crisis. In courses from biology to psychology, it became clear that the energy system was the root of the problem. I tried to figure out ways to get more people to change how they used energy. Working with other students and faculty, we had thousands of students and employees take part in an experiment. They were trained in why saving energy was important and asked to do things like take the stairs rather than the elevator. Some became zealous, switching off bathroom lights while people were still on the toilet. The campaign worked: We helped university residences cut around 10 percent of their energy use. Before I graduated, I helped build on campus a large solar project, whose profits would fund student scholarships. It was clear to me that changing the energy system was the key to tackling the climate crisis. And I vowed to do everything I could to help drive that transformation.

Yet the results were not satisfying enough. I wanted a bigger scale. I saw that changing behaviors was not as powerful as changing institutions. So I spent the next decade trying to understand how people have tried to change energy policy to tackle the climate crisis. I tell you these stories to show you how—to paraphrase Rilke—I've lived my life in widening circles. How I've tried with each passing year to reach a bit further into the energy system and drive greater changes.

I tell you these stories because I know: You can do the same.

The climate crisis is here now. How have we gotten into this mess?

One common answer is that *we are all to blame*, through our everyday choices: whether to drive, bike, or walk, to take a flight, or to buy stuff. It's easy to see why we might think that environmental problems, including climate change, are mostly a question of our own behavior.

When we buy something, we are making decisions that shape the future. It's easy to get lost in the smaller choices, like using a plastic straw. We can forget that the bigger choices, like what car we buy or whether to buy a car at all, have a much larger impact. It's even easier to miss the choices we can't make, like taking a train to a nearby city when no train service exists.

When we start to see the choices that are *not* available, we can begin to see the role of political power in our daily lives. Who decides what options are available for us to choose in the first place?

Institutions shape the choices we *can* make. Some actors in society have more power than others to decide how our economy is built and fueled.

When it comes to the energy system, fossil fuel companies and electric utilities have limited our choices. Over the last century, these companies resisted innovation. They learned how to do one thing well—dig up, sell, and burn fossil fuels—and they want to keep doing it. When electric lightbulbs were first invented in the late 1800s, the new industry was radically innovative. Thomas Edison, for example, hoped to move away from coal toward wind and solar energy. But by the 1920s, this inventiveness declined as corporate managers took over. They focused on existing technologies, building ever-larger coal plants. And they worked to undermine more efficient technologies, which would have enabled the energy system to produce less waste.

Stifling innovation wasn't the only way these companies played a leading role in creating the climate crisis. Fossil fuel corporations and electric utilities also organized to deny climate science. The steady drumbeat of this well-funded campaign broke through. When I think back to my first introduction to climate change in high school, I often wonder why my otherwise brilliant teacher thought to present the topic as a debate. Well, at the time, fossil fuel companies and electric utilities were a decade into their campaign to discredit climate scientists. From the 1980s to the present, they have funded organizations that denied the scientific consensus on climate change, spending billions on the effort. And these efforts were wildly successful: The exact language used in denial reports permeated the public discourse, ending up in the media and in presidential speeches.

Throughout the 1990s and the first decade of this century, clean-energy advocates fought back. They managed to get laws on the books to ramp up renewables in a majority of states. But just as we started making progress on cleaning up our energy system's dirty ways, these corporations launched an assault on climate policies. As I document in my book, *Short Circuiting Policy*, electric utilities and fossil fuel companies have spent the past decade attacking the piecemeal climate policy we have managed to implement so far.

While many states have clean-energy goals, overall the pace of change in the electricity system across the country is paltry. In 2018 only 36 percent of the U.S. electricity supply came from clean-energy sources, including wind, solar, hydropower, and nuclear. Between 2009 and 2018, annual growth in renewable energy was a mere 0.7 percentage points. Meanwhile, natural gas expansion has outpaced renewables in recent years. Incremental progress—with one step forward and another back—simply won't get us to 100 percent clean electricity in the coming decades.

Our challenge is only compounded when we realize that we also need to grow the grid to power our homes, businesses, and cars with clean electricity. Of course, if we invest in energy efficiency, we won't have to build as much new energy supply, because efficiency reduces overall demand. That's especially important to do today, when our energy system is still powered by dirty fossil energy: If we use less *now*, we will burn less fossil fuel overall. And if we keep safe nuclear plants open for as long as possible, it will also be easier to make progress—they currently supply more than half of our clean power. Right now, when a nuclear plant retires, it is often replaced with dirty energy. No matter how you slice it, we will need to move a lot faster to change our energy system.

But rather than helping drive the changes we need, electric utilities and fossil fuel companies have blocked clean-energy laws from passing in new places, leaving many states without any target. They have also worked to roll back or weaken clean energy laws on the books. When these strategies failed, electric utilities and other polluting companies made the fight bigger, using the public, political parties, and the courts to short-circuit progress on climate policy. As a result of this deliberate delay, most states are way behind on cleaning up their energy systems.

By bending the political system to their will, fossil fuel companies and electric utilities have created the climate crisis. Big, polluting corporations have severely limited the choices we can make today. We are not all equally to blame.

Once we see that the climate problem is an energy problem, and that our energy problems are institutional, political, and coordinated, we can also see that climate action must take the same form.

Take, for example, "flight shaming"—the movement to use social pressure to discourage people from flying. Likely you know that you "shouldn't" be flying because it takes a lot of fossil fuels to do it. If you fly, and you've ever tried using an online carbon footprint calculator, you've probably found out that flying is the biggest slice of your personal carbon emissions by far.

If you live on the East Coast, where there is a train service between major cities, maybe you tried to take the train instead. Other times, you may have decided not to take the trip.

But you may have family in another country or need to take a plane for your job. Maybe you've bought some offsets and tried to reduce your emissions elsewhere, to lessen your guilt.

Yet when is the last time someone tried to make you feel bad for paying taxes and contributing to your country's military budget? I'd bet this hasn't happened. The U.S. military is an energy guzzler—it is the institution that consumes the most fossil fuels in the world. In fact, its carbon emissions are greater than those of many countries. In recent years, the U.S. military has emitted more greenhouse gases than Denmark.

The military's annual emissions are similar to those from commercial aviation. Yet those individual footprint calculators don't consider this part of your personal footprint. There's no button to click to offset your contribution to the military's emissions.

Nor can you easily calculate or offset the emissions from the steel that was used to build the office where you work. And it would be tough to offset the fertilizer—made from fossil fuels—that grows the food you eat. The more you look, the more you see how fossil fuels are everywhere.

On average, each American emits around fifteen metric tons of carbon pollution each year. You could work very hard to try to shrink your own tonnage but then wake up one day to realize that the United States as a whole emits around six billion metric tons every year.

No one can unilaterally choose to live in a low-carbon economy.

The goal is not self-purification but structural change. As Bill McKibben has put it: "Changing the system, not perfecting our own lives, is the point. 'Hypocrisy' is the price of admission in this battle."

A simple thought experiment can help illuminate this point: If you were to die tomorrow, how much less carbon emissions would occur? Not much less. Society would keep churning along, belching out pollution as it went.

And this isn't even hypothetical: We ran the experiment in 2020, during the coronavirus pandemic. For months few were flying or driving or even leaving their homes. Yet emissions barely budged—falling an estimated 8 percent. We need emissions to fall that amount every year until 2030 to limit warming to 1.5 degrees Celsius. Individual action alone won't get us there.

But what if you build or contribute to something bigger than yourself, something that breaks the relationship between energy and pollution in the longer term? That would keep going beyond your lifetime.

Put simply: We cannot make enough headway on the climate problem by working at the individual level. We need to organize our efforts. And that is one essential function of a modern, healthy democracy: cooperation and coordination.

When Greta Thunberg was asked to name the top thing that people could do, she answered: "Try to push for a political movement that doesn't exist. Because the politics needed to fix this doesn't exist today. So I think what we should do as individuals is to use the power of democracy to make our voices heard and to make sure that the people in power cannot continue to ignore this."

And we won't move fast enough if we don't draw in everyone. We need celebrities with private jets and people who have never taken a flight. We need meat-eaters and vegans. We need people living in West Virginia—the state with the second-dirtiest electricity mix—and people living in California. The pace and scale of the changes we must make in our energy system are unprecedented. We need everyone in the fight.

When you decide to become part of changing our energy system, you can start small. In this first circle, you're training yourself to see the energy system and how to change it.

It's great to turn off the lights when you leave a room, and better still to stick up a sign to remind others to do the same. Or announce out loud before you leave a room, "Could the last one out hit the lights?" This kind of action can help conserve energy at the margins.

The challenge with relying on ongoing behavior change is that people must make the choice. Sometimes people simply forget. Hence, it's even more effective to aim for changes that are structural and hard to reverse.

The building you live in is likely fueled by natural gas. We use it to cook our food, heat our water, and warm our rooms. This fossil gas is methane, which creates massive amounts of radioactive waste when it is dug up from the earth. When you use it in your home, it can leak and make you sick. In rare cases, it can even cause an explosion. Leaking fossil gas is not just bad for your family; it also heats the planet far more powerfully—roughly thirty times more over a hundred-year period—than carbon dioxide.

But it doesn't have to be this way. We can use clean electricity to power induction stovetops, electric water heaters, and electric furnaces and heat pumps.

So if you want to change something in your daily life, make your home all electric. Start planning for that change now: Save up, look for a contractor, see if your city offers any incentives. If you don't own your home, you can talk to your landlord, friends, parents, or grandparents about making this change.

And if you do manage to rid your home of fossil gas, consider sticking a sign on your lawn that says: "Ask me about my all-electric home!" Talk to your friends and neighbors about what you did and why you did it. And try lobbying your city, county, or state to offer financial programs to support more people in making the same choice you did.

Changing the energy system can start in your own life. But it's important to remember that it cannot end there. You have to keep looking for the wider circle.

The next circle is found in community. Each one of us is connected to others. These relationships give us more power than we believe. We can and do shift our communities when we act.

We can begin by talking about the climate crisis with others. This may feel small, but in truth, it's big. So many Americans barely hear about the weather systems shifting perilously around them, much less solutions to address the problem. As Katharine Hayhoe has said, "The number one thing we can do is the exact thing that we're not doing: talk about it."

I try to talk about climate change once a day, with someone other than my cat. Often I'll begin a conversation about the weather—commenting on a drought or heat waves—and before you know it, we're talking climate crisis.

There are a range of reactions to my raising the topic.

One man, whom I talked to in a cab, had heard of the problem but thought that volcanoes were the cause. I explained to him that fossil fuels were the culprit, and by the time I left, he understood our predicament better.

Another man, who was older and didn't have kids, commented that he never does well in the heat. I told him it would likely keep getting worse from here on out. He said that was true. But he wasn't worried about it—because before things got too bad, he would be dead. This kind of reaction is a privilege. To divorce yourself from the ongoing struggles of humankind. To not worry about what will befall whom, and when, because you will be long gone.

It's also a sign that someone is trying to protect themselves psychologically. Many people want to hold the climate crisis at a distance so they don't have to feel sadness or fear. They don't want to reckon with it.

As Wendell Berry once put it: "It is the destruction of the world in our own lives that drives us half insane, and more than half."

Given the emotions that this conversation can raise in people, it's important that you talk about solutions as well. Remind people that our governments—city, state, and federal—can make different choices. That this story is not yet finished.

These days I don't just talk to people one on one about how our energy system is driving the climate crisis. I talk to journalists, go on podcasts, give talks in churches, at universities, to politicians. I do as many talks as I can, to whoever wants to hear from me. Because every conversation is a chance to wake our democracy up.

As you continue to work in the community, it can be helpful to join an organization. These groups can help you understand what next steps you can take to advocate for climate action. There are lots of great groups to consider joining, including 350.org, Citizens' Climate Lobby, Greenpeace, Sierra Club, Sunrise Movement, and Surfrider Foundation. Try to find one with a local chapter in your community. In some places, there are also regional groups that focus on environmental justice, like WE ACT in New York, Dogwood Alliance in the southern states, and the California Environmental Justice Alliance. You can also give money to these groups, or others, to help them fund their work.

Linking up with organizations can help you extend your reach. As Jane Fonda, who started Fire Drill Fridays, has said, "People who are organized can change policy. People who are together, unified, and organized around a strategic goal can change anything."

We are working toward the widest circle, which is policy change. It's not easy, and you can't do it alone. But each one of us can chip away at the laws that keep us stuck in our current energy system. Slowly, working together, we can shape it into a new form.

As part of my academic research career, I've had the privilege of interviewing more than 150 leaders who have worked to change energy policy. The people I admire the most are the ones who have used their ideas to try to change the energy system. One woman, Nancy Rader, wrote a master's thesis in the early 1990s on an idea she called a "renewables portfolio standard." With effort from a lot of other brilliant activists across the country, that policy now exists in most states. Mary Anne Hitt and her team at the Sierra Club had an idea to stop new coal-fired plants and shut down existing sites in the United States.

The program she directs has stopped more than two hundred plants from being built and helped retire more than three hundred others. After being harassed by fossil fuel companies, Naomi Oreskes began investigating them. Her research has resulted in lawsuits against major oil companies to try to hold them accountable for climate denial. Alexandria Ocasio-Cortez wasn't yet sworn in when she joined young activists, including Varshini Prakash of Sunrise Movement, for a sit-in on Capitol Hill and demanded a Green New Deal. Less than a year later, all the major Democratic candidates running for president had backed the policy idea.

Of course, none of these women are changing policy alone. They have coauthors and colleagues, funders and friends. But each one of them has become a linchpin in the effort to transform our energy system—they are people doing what matters. And because they are working in a wide circle, their actions can change our energy system, not just their own lives.

What policies could you try to work on? You could try to stop new fossil fuel developments in your state or work to shut down coal and natural gas plants that already exist. You could try to support more renewable energy projects by backing policies that pay people and companies to build them. You could support state and federal policies that speed up electric-vehicle adoption by buying back old clunkers, giving funding for new cars, or investing in charging infrastructure. On the local scale, you could work to support public transit or protected bike lanes. You could pressure your city to ban natural gas in new buildings, as cities across the country are starting to do. And you could work to get your state government to provide financial incentives for people who electrify their homes or increase their energy efficiency. To hold electric utilities accountable, you could try to get your state to pass an "intervenor compensation program," which pays advocates to show up at regulatory proceedings to speak in the public interest. And you could call your federal representative to say you want more funding for research and development, including the Advanced Research Projects Agency–Energy (ARPA-E) program.

To figure out what policy you'd like to work on, lean on the organizations you've joined. See if they have campaigns you can plug into or if there are ways you can start your own and get others to join you.

Until policies are in place that effectively challenge fossil fuel companies' and electric utilities' political dominance, the lives of billions of people, communities, species, and ecosystems are in grave danger.

The twentieth century will be remembered as an age of destruction and delusion. It will be remembered as a time when we filled our oceans with plastic, our lungs with poison, and our minds with climate denial.

With effort from all of us, the twenty-first century can be an era of healing. We can spend this century reducing carbon pollution, bending the emissions curve downward. We can put policies in place that make it easier for our friends and neighbors—and people across the oceans and in future generations—to live without pollution.

This will not be easy. It will require us to get up every day and chip away at the problem. To remind myself, I keep these words from Mary Oliver running through my head: "May I be the tiniest nail in the house of the universe, tiny but useful." Each one of us can be that nail, chipping away at the fossil energy system.

Do not demand that your smallest, personal circle be pure before you start working on the broader circles of community and policy. Because that day will never come. Let's dig in today to shift the system—and tomorrow and the day after.

When I come to the end of my life, I want the scales to show that I prevented more carbon emissions than I caused. And there is no way to make that happen if I work only on myself.

My offset plan is activism.

Mornings at Blackwater

MARY OLIVER

For years, every morning, I drank
from Blackwater Pond.
It was flavored with oak leaves and also, no doubt,
the feet of ducks.

And always it assuaged me
from the dry bowl of the very far past.

What I want to say is
that the past is the past,
and the present is what your life is,
and you are capable
of choosing what that will be,
darling citizen.

So come to the pond,
or the river of your imagination,
or the harbor of your longing,

and put your lips to the world.
And live
your life.

Like the Monarch

Sarah Stillman

As an investigative journalist, I'm prone to accumulating large piles of useless paper on my kitchen table: press releases, public records, half-full notebooks. Sometimes I go months, or years, without a purge. This winter, when I attempted a big one, I noticed a thick file from a story that haunted me in the wake of Hurricane Katrina. The folder was labeled "Signal." I figured I'd toss it; the hundreds of records stemmed from a civil lawsuit that seemed dated. Then I read one: a legal declaration from a man named Sony Sulekha.

As I read Sulekha's words, the details rushed back. The lawsuit involved one of the largest labor-trafficking cases in U.S. history. It was anchored in the claims of 590 men, many of whom had come from India in late 2006 to help rebuild Gulf Coast oil rigs after the storm. Sulekha was one. He'd convinced his wife to sell her jewelry and the family land so he could provide their household a better future. That's not how it worked out when he arrived in Pascagoula, Mississippi. There he and his colleagues faced what the suit describes as fraud, coercion, assault, false imprisonment, "intentional infliction of emotional distress," and more, at the hands of an Alabama-based marine construction firm called Signal International.

Reading Sulekha's testimony felt like uncorking a message in a bottle from the post-Katrina coast, addressed to the world we now have to navigate, this landscape of superstorms and annual "hundred-year" floods and forest fires, which is transforming not just our environment but also the nature of human migration. I spent the night going through the folder. It was true to its name: a signal, a cautionary tale.

The scale of human dislocation yet to come from the climate crisis verges on unimaginable. One starting point is internal displacement; the vast majority of climate migrants will relocate without crossing national borders. Wildfires, hurricanes, heat waves—in 2016 "sudden-onset" disasters pushed three times as many people from their homes as conflict or violence, according to the Internal Displacement Monitoring Centre. A World Bank report estimates that, over the next thirty years, 143 million people will be displaced within three of the most vulnerable regions alone: sub-Saharan Africa, South Asia, and Latin America.

Aid workers often call these individuals "IDPs," for "internally displaced persons." But I've yet to meet a single person, across a decade of covering migration issues, who sees herself or himself in this acronym. People define themselves instead by the kinds of fruit trees they grew in their backyards before a hurricane came, or else by the ways their families worked the land until a drought rendered it too dry for them to stay. A conspicuous feature of life as an IDP is uncertainty: Will you be away from home for days, or decades? In the former group, I think of my brother, who swaddled his newborn son and four-year-old daughter and carried them north in 2018 to avoid the smoke of California's forest fires, which had crept all the way into San Francisco. His return was fast and simple, as he'd known it would be. But increasingly, families flee homes, or entire regions, to which they might not have the choice of return.

The same is true for the second, and contested, group of climate-displaced people: Sometimes referred to as "climate refugees," they cross international borders on account of a climate factor. Since 2008 some 25 million people have been displaced globally, often permanently, by catastrophic weather events each year. Others have left their countries on account of "slow-onset" changes that verge on existential—desertification, rising sea levels, land degradation, persistent drought. Already some of these families have sought relief at the U.S.-Mexico border, where the Trump administration has tried to build a wall, in every sense of the term.

Who qualifies as a refugee is a matter of debate. Under international law, the answer looks narrow, even antiquated. The UN's 1951 Refugee Convention first defined the protected legal category, in part, to cover those who faced a "well-founded fear" of persecution, on account of threats tied to five particular grounds: "race, religion, nationality, membership of a particular social group, or political opinion." These sanctioned categories fit the world of threats perceived at the time by the convention's authors. But for many modern families, the status is a safety net full of holes, offering uncertain protections for those who've fled gang violence, gender-based violence, or other less tidy threats. Protection for asylum seekers hinges, in U.S. immigration court, on arbitrary factors like the jurisdiction—or even the courtroom—in which a case is heard. For those who've fled climate change, current law has little to offer.

Modernizing the legal construction of the "refugee" is an option on the table to make the grounds for protection more capacious, or at least more reflective of our century's threats. But the political winds have shifted in the other direction, making attempts to renegotiate the Refugee Convention's terms a significant gamble; populist leaders are rising globally, with an agenda of contraction—building walls, slashing refugee resettlement numbers, eviscerating the post–World War II commitment to *non-refoulement*. (The term is French for "nonreturn" and reflects the Refugee Convention signatories' commitment not to repatriate people to countries where they face likely death or persecution.) So-called climate refugees often come from countries that have contributed far less to the causes of climate change than their wealthier counterparts. Still, the limitations of our current legal frameworks ensure they pay a disproportionate price.

Some migrant families have fought to win recognition as "climate refugees" in court. One prominent example is Ioane Teitiota, who left the Republic of Kiribati, in the Pacific, to seek refugee status in New Zealand. He argued that rising sea levels and other climate-change effects in his home country violated his right to life. After New Zealand rejected his claims and deported him and his family in 2015, Teitiota filed a case with the UN Human Rights Committee. This January, the panel made a landmark ruling: Although New Zealand's courts didn't violate Teitiota's rights in his specific case (in part because Kiribati is still considered ten to fifteen years away from uninhabitability), climate change did, in fact, offer potential grounds for asylum in future cases.

For those whose circumstances don't fit easily within the categories of "internally displaced persons" and "refugees," we need a third category. Millions of people's migration stories are being rewritten by climate change in ways that aren't reducible to narratives of escape. Sulekha and his Indian guest-worker colleagues at Signal International belong to one such group. The decades ahead will see countless immigrant workers help rebuild communities when their environment changes or extreme weather strikes.

Sony Sulekha planned to do just that in Pascagoula, Mississippi. Back in 2005, he saw an ad in the local newspaper seeking skilled metalworkers for jobs in the United States. Sulekha was intrigued. Traveling more than two hundred miles from his home in Kerala, India, he attended a recruitment session in the city of Cochin. He checked out Signal's website, where he saw photos of well-appointed living quarters: "a nice room with a washing machine, refrigerator, and microwave oven." Excited, he took out loans to pay the steep recruitment fee charged by labor brokers: more than $13,000.

On November 19, 2006, Sulekha left Mumbai around midnight. He flew with thirty-some fellow workers to New York and then Atlanta and then Mobile, Alabama. There the men boarded a bus bound for Mississippi.

Pascagoula, too, might have pleased Sulekha, if he'd been allowed to explore it. The town had resorts and bike trails and beaches, as well as a history of civil rights protests that might have moved him. But when a Signal International escort brought Sulekha to his new living quarters, he felt shock, according to his legal declaration. "I lost my breath," he stated. Twenty-four men were bunked in a single room with two toilets. "I had never seen such horrible accommodations," he said. The men's living quarters were surrounded by barbed-wire fencing. "I felt," Sulekha said, "like I was in jail."

The job itself proved risky. Sulekha was exposed to industrial dust, which made him sick. When he complained, he learned he had no access to proper medical care—although the company, he noticed, routinely deducted money from his paycheck for "healthcare." Signal charged him $1,050 a month to live in the guarded labor camp surrounded by barbed-wire fencing, where he described feeling "very isolated" and "increasingly depressed."

Sulekha's colleagues in Pascagoula had similar stories. So did hundreds of Indian guest workers at another Signal site, in Orange, Texas.

These men were telling a climate story. Mostly, we didn't know how to hear it.

Hurricane Katrina gave the country a traumatic preview of large-scale climate displacement. The storm damaged more than 100,000 housing units in New Orleans and pushed at least 800,000 people out of their residences. This was not equal-opportunity displacement. In New Orleans, Black families of low socioeconomic status bore the brunt of long-term dislocation.

Years after the storm, the Institute for Women's Policy Research interviewed 184 Black women who'd lived in four of the city's largest public housing complexes. The majority reported wanting to return to their homes but being unable to do so because city and federal officials had demolished the buildings. The women had been pushed into more expensive housing, sometimes in neighborhoods where they faced racial intimidation. A decade after the storm, nearly four in five White residents of New Orleans said their state had "mostly recovered," while three in five Black residents reported it hadn't, according to Louisiana State University's Public Policy Research Lab.

Another toll of Katrina was a massive labor shortage. Across the Gulf Coast, the oil industry scrambled to rebuild. Signal International had experience repairing offshore oil-drilling rigs in states like Mississippi and Texas; the company realized that profits would soar if it could help get oil companies back to work efficiently. To do so, it would need skilled metalworkers—many and fast. So it turned to the H-2B guest-worker program, which helps companies outside the agriculture industry bring in immigrant workers from overseas. But this program also rendered migrant workers vulnerable. Their visas yoked them to a single employer, without whom they'd lack legal status.

One result of this ill-regulated arrangement was exploitation. On-site injuries, verbal abuse, wage theft—all were common. This wasn't true just for Signal International. "After Katrina, every large-scale employer was competing for workers," Saket Soni, who at the time directed the New Orleans Workers' Center for Racial Justice, told me. But recruited workers, he said, were treated as commodities: "They weren't involved in, or empowered by, the terms of arrival."

Beyond guest workers, all kinds of immigrant workers served the

Gulf Coast's post-Katrina recovery, including undocumented workers and some asylum seekers. After more recent disasters, immigrant construction crews have proven indispensable. Lin-Manuel Miranda, in *Hamilton*, sums it up well: "Immigrants: we get the job done."

Recently, Saket Soni met with workers in the Florida Panhandle who were helping to rebuild after Hurricane Michael. He told me that 90 percent of the workers fixing damaged roofs and the like were undocumented. "Many of those workers are the same people who rebuilt New Orleans after Katrina, then Baton Rouge after Gustav, then Houston after Harvey and North Carolina after Hurricane Florence," he said. "We have a new itinerant workforce, almost like the Irish travelers—a roaming set of construction workers." He added, "It's safe to say that the number of Latino immigrants and undocumented immigrants is far more concentrated in disaster-hit and climate-hit areas."

These days, many recovery workers come from countries where climate change has left its mark. In the years since Sulekha left Kerala, India, bound for Mississippi, his hometown has been hit by severe monsoons and droughts alike. In 2018 Kerala endured the flood of a century, followed a year later by a monsoon that brought more lethal flooding and landslides, displacing some 200,000 people. The chaos spurred a government debate about whether climate change played a role and about the need for resiliency planning. "The monsoon calendar has changed and its intensity too," an official from the India Meteorological Department told *India Today*.

Many locals will stay put as already-thin margins of survival narrow. Some will invest their sweat in adaptation—reforestation efforts, flood mitigation, the building of temporary storm-time shelters. But others, inevitably, will leave, for a jumble of reasons of which weather is one. "Economic migrants," the world will call them, placing them on the "undeserving" side of an arbitrary line. At a migrant shelter in the Rio Grande Valley, in Texas, I recently met a South Indian teenager who had just crossed the border, unauthorized and alone, wearing a Rosie the Riveter T-shirt. He told me that his family back home had paid nearly $40,000 for his journey, accruing debt in the hope that their son could provide a lifeline through remittances. Like most migrants, he didn't have one simple reason for the journey, which was about to lead him to a gas-station job in Indiana. Food insecurity was a factor in a complex constellation exacerbated by climate change.

This is true for hundreds of thousands of migrants who've arrived at the U.S. southern border in recent years. The vast majority have come from Central America's "Northern Triangle"—Guatemala, Honduras, and El Salvador—where inhabitants of the region's "dry corridor" endure increasingly erratic weather patterns, including droughts and heavy rain. Gang violence and corruption lead the list of reasons that families leave Honduras and El Salvador to seek asylum. But last year in Honduras, a drought compounded instability, prompting the government to declare a state of emergency. In the western highlands of Guatemala, climate change is now considered a major catalyst for migration. Adverse weather conditions have devastated crops, leading to widespread food insecurity; according to the World Food Programme, nearly 50 percent of Guatemalan kids under the age of five are considered chronically malnourished, with rates exceeding 70 percent in many rural areas. Last year, more than 1 percent of Guatemala's entire population attempted to reach the United States.

Speaking to asylum seekers at the southern border and elsewhere, I've noticed: Climate migration is rarely reducible to a single variable. Recently I spoke to a young mother in Mogadishu, Somalia, whose husband, Ahmed Salah, had presented himself at a port of entry in South Texas to seek asylum. His problem's roots were gnarled: He'd fled conflict in Somalia, where drought has fueled food scarcity, which, in turn, has worsened armed conflict. He'd sought safety elsewhere in Africa, but, finding little of it, he'd journeyed thousands of miles across Latin America, trekking to the United States as his best hope of freedom. He was apprehended and detained. Eventually he went before a judge in Louisiana, who had a 100 percent asylum denial rate: All day in court she rejected each case before her. Salah proved no different; the judge rejected his asylum claim and sent him back to Somalia. There, last March, he died in a car bombing.

Climate change doesn't just spur migration; it makes preexisting migrant communities more exposed to extreme weather. As director of the Global Migration Project at Columbia Journalism School, I worked with a team of three postgraduate fellows to document this trend. We spoke to more than two hundred climate scientists, migration experts, legal scholars, and migrant communities on five continents. Some of the reasons we found for this disproportionate impact were obvious. From Bosnia to Kenya to Bangladesh, refugees tend to live in precarious housing, like tent cities and plywood shacks and makeshift camps; they're often at the mercy of the elements as a result. Undocumented communities may decline to seek help from authorities when a storm or fire strikes, fearing arrest or deportation. The same is true of postdisaster aid.

But elsewhere we found surprises. The Global Migration Project team traveled to the Bahamas after Hurricane Dorian struck in September 2019. Amid the devastation in the Abaco Islands, we learned about a large Haitian migrant population living in so-called shantytowns ravaged by the record-breaking storm. In the aftermath, the government promised Haitian migrants shelter and aid, free of repercussions. But within a matter of weeks, immigration authorities began a campaign to round up, arrest, detain, and deport hundreds of undocumented Haitian storm survivors, sending them back to a country facing political unrest and its own climate instability. Many women alleged sexual assault and verbal abuse in the process. Some were pregnant or nursing or caring for small children who'd been born in the Bahamas but who were deported nonetheless to Haiti, a country they'd never seen. We are not just living in the age of the climate refugee. We're also witnessing the advent of the climate deportee.

To face a threat of this scale, we need new frameworks—legal, cultural, political, linguistic. "Language alone," Toni Morrison said in her 1993 Nobel lecture, "protects us from the scariness of things with no names." But where do we find the words?

Dominant groups have long used menacing metaphors from the natural world to frame migration debates. Swarms, surges, hordes—animal allusions offer nouns of choice to degrade and belittle, to rationalize subjugation and criminalize survival. President Trump rails against the "catch and release" of asylum seekers at the southern border, as if they were fish; his White House denounces, in press releases, "the violent animals of MS-13."

This long history of dehumanization makes me reluctant to seek lessons about human migration from nature. But the animal world has more to offer than an appendix of denigration. Consider monarch butterflies, the ultimate seasonal migrants. For years they've offered a visual language to the immigrant-rights movement. Monarchs travel more than two thousand miles annually, from Mexico to Canada, using the angle of the light as their guide. They've proven the possibility of unfettered but organized migration. "Like the monarch butterfly," notes the artist Favianna Rodriguez, "human beings cross borders in order to survive." When traveling Mexico's migration trail with Central American families, I've spotted monarchs painted on the walls of migrant shelters and graffitied near the train tracks of "La Bestia," the rails that migrants and asylum seekers ride.

But monarch migrations in recent years have also served as a warning. The butterflies are exquisitely good at climate adaptation, on account of their large geographic range. But their sensitivity to temperature has also made their routes, and their reproduction, unstable amid changing weather patterns. The World Wildlife Fund reports the monarch population is in steep decline; over the past twenty years, the butterflies have lost more than 165 million acres of habitat. The transnational migration of some monarch populations is now classified as "endangered."

Whales present a different climate-displacement story. In the summer of 2018, I spent time in McAllen, Texas, covering family separations at the border. At a federal courthouse, I watched parents who had been pulled from their children face criminal prosecution, wearing shackles, for "illegal entry."

When I left, I sought solace on Orcas Island in the Pacific Northwest, where my parents live. My dad, in July, likes to sit on the shaded deck in his oak-wood chair, hoping for a rare sighting of the resident killer whales, the "J Pod," a family of more than twenty whales. The pod has long made the San Juan Islands its summer home, a stop in the broader waters of the Salish Sea. (In the winter months, the Southern Resident whales hunt for salmon farther south, sometimes as far as Monterey, California.) But in recent years, sightings have grown rarer. The Southern Residents, an endangered population, are no longer able to find Chinook salmon to eat, a problem at the top of a long list of interrelated threats: boat traffic, toxins, warming waters. So they're shifting their travels, in search of food and ease. Animal or human, that's migration's basic compass: survival.

In the waters just beyond my parents' home, a mother whale was in mourning. Tahlequah, as she is known to those who study her, is a member of the J Pod. On July 24, Tahlequah gave birth to a calf, who died within half an hour. The death reflected a broader social crisis; in the prior three years, the Southern Resident whales hadn't had a single successful birth. After the loss, Tahlequah carried her lifeless calf through the water. The ritual lasted a day, and then another. As a week passed, the story became an international drama.

One San Juan resident reported seeing Tahlequah surrounded by six other female orcas in the hours after the calf's death, moving in a tight-knit circle under the moon. The whales of J Pod then traveled alongside Tahlequah and her dead calf for weeks. When she got tired, other whales took over the task for her. Tahlequah continued this way for seventeen days and more than a thousand miles.

Killer whales, I learned meanwhile, are matrilineal. Grandmothers are vital to the group's survival; according to a study in the *Proceedings of the National Academy of Sciences*, the pod turns to the eldest female whales to lead their foraging efforts, a particularly vital skill when food is scarce. These postmenopausal females draw on stores of "ecological knowledge" derived from decades of life experience. The whole pod arrives at strategies together, turning to their culture and history for clues, and picks up the slack for one another, sharing food.

The J Pod's response to loss is one model for meeting the fear the climate crisis can inspire, and the grief: through acts of public witness, individual and collective. Humans, too, have a gift for this.

In Pascagoula, the Signal guest workers refused to stay cowed behind barbed wire. Sony Sulekha and others gathered quietly at informal meetings to discuss their poor wages, housing, and food. When the men lodged complaints, a Signal boss said, "Indians are like animals." Signal employees threatened the men with deportation.

Tensions grew as the guest workers organized. Company personnel reached out to U.S. Immigration and Customs Enforcement, seeking guidance on how to fire "chronic whiners." In March 2007, Signal's private security guards staged a predawn raid, seizing several of the workers who'd taken leadership roles and detaining them. One, Sabulal Vijayan, felt terror at the prospect of deportation; he cut his wrists, trying to kill himself.

Sulekha and some of his colleagues went to the hospital with Vijayan to look out for him. Vijayan had been a leader among the group, and when he healed, the men took their protests public. Eventually, with the help of the New Orleans Workers' Center for Racial Justice, they staged a monthlong hunger strike in Washington, DC. "We were like pigs," Vijayan told the press, "in a cage."

Both Sulekha and Vijayan also came forward in a civil lawsuit, one of the largest human-trafficking cases on record, which made claims under the Civil Rights Act of 1866 and other provisions. High-powered groups—including the Southern Poverty Law Center, the American Civil Liberties Union, and a handful of prestigious law firms—helped represent the workers pro bono. Accountability took a village. Signal fought to get the lawsuit thrown out. Instead, a judge let the case move forward.

Reading through my Signal file from the first of several civil suits, I thought of Indra's net, a notion born of Hindu cosmology. The image is this: a net within which every jewel is tied to every other jewel; each jewel, in turn, reflects all the others, a portrait of interconnectedness, wherein no one suffers in isolation nor rises alone. Indra, in Hindu teachings, used the net to ensnare enemies. Few strengths rival a tightly woven social web.

So many of the hopeful responses to Hurricane Katrina, and the storms that followed, affirmed that idea. The most vital rebuilding has come from frontline communities strengthening not just their physical structures but their social ones. The Gulf Coast Center for Law & Policy, led by Colette Pichon Battle, works, for instance, to "advance structural shifts toward climate justice and ecological equity." Many in the Deep South are building plans for migration with dignity within communities where sea level rise now threatens ways of life. They are working, too, to hold accountable the primary forces responsible for climate change, including the oil and gas industries.

In 2015 the trafficked guest workers of Signal International won a victory against one small part of this industry. After a four-week trial, a jury in New Orleans ruled against Signal as well as its codefendants, and in the guest workers' favor, awarding the plaintiffs $14 million in damages. The company then settled eleven more cases on behalf of hundreds of additional Signal guest workers from India for an additional $20 million. Signal declared bankruptcy and offered a formal apology. "Signal was wrong," the company said, "in failing to ensure that the guest workers were treated with the respect and dignity they deserved."

At my kitchen counter, I decided to keep my Signal folder. I think of the workers' stories often amid national cycles of flooding and burning and rebuilding. I see Sony Sulekha leaving Kerala to fly west toward the wreckage of Hurricane Katrina, flush with hope, even as families from New Orleans trekked north, their lives temporarily reduced to duffel bags. I see him rushing to the hospital to join his suicidal colleague after an immigration raid conducted in retaliation for workers' organizing—then later chanting with that same colleague in the street, sharing their story with the public. The whole group comes to mind in the courtroom, enumerating and asserting their legal rights and putting an oil-rig-repair company out of business in the process.

Recently, the Signal workers' experience was one of the many factors that spurred Saket Soni and his colleagues to build a new organization, Resilience Force. It aims squarely at the disaster-rebuilding challenge ahead. "The U.S., right now, is completely unprepared for the scale of migration that climate change will bring—both internal migration and the arriving climate refugee from across the border," Soni told me. "At the community level, none of this is inevitable. Communities can plan for climate change and displacement, and they can have a view on when and how and where they want to go." Resilience Force is helping to shape those proactive plans, he said, and to organize for a "more effective and equitable approach to disaster preparation, response, recovery, and rebuilding."

Soni's network hopes to imagine the broadest possible coalition to do the work of ensuring climate resilience, drawing on the lessons of Katrina. Indigenous communities, migrant workers, workers with deep roots in their regions, all kinds of families striving to rebuild their homes or relocate—they hope to stitch a net that's vast enough to hold, and reflect, them all.

In our time of disturbance and radical change, we are crossing a threshold, a portal, or an unseen bridge from one world to another. It could be said that the bridge is either collapsing beneath us, or being made as we walk together, in the long twilight hours when one civilization gives way to another.

—Geneen Marie Haugen

Community Is Our Best Chance

Christine E. Nieves Rodriguez

Two days before Hurricane Maria hit Puerto Rico, I was there, waiting in line at the gas station. The line wrapped around the street with no end in sight. At least this gas station had an usher who was directing people to each pump. I called my mom as I waited and told her I was getting ready. Her voice was shaky. She knew this hurricane was going to be catastrophic. Instead of telling me how afraid she was, she asked me: Why was I wasting my time and education in Puerto Rico? Why had I decided to come back after working so hard to get out and make something of myself?

I hung up without an answer. Tears ran down my face as I inched my way to the pump. I wish I had known how right she was to be afraid. But then I didn't know what was coming. I kept wondering: Had I made a mistake after all?

I left Puerto Rico when I was eighteen because I was convinced there was no future for me here. After more than a century of colonial status to the United States, Puerto Rico was far from an enchanted island for me. I didn't see the beauty; I only felt bitterness. Bitterness toward the corrupt government, bitterness toward the elite families and corporations that seemed perfectly comfortable in the extreme inequality. Bitterness toward the trash on the beaches and streets, the constant defacing of historical sites, the poor maintenance of public areas. That disdain became a mantra that everyone else was also echoing—that we are lazy and corrupt, that nothing good can happen here. So I left with the same feeling many Puerto Ricans had left feeling—that we were inferior to the Western world's superiority. And I was determined to prove my worth, to myself, to Puerto Rico, and to the world.

I wanted to make a difference—to use information to open up the world for others—because I had felt so trapped. I pursued an Ivy League education (though I didn't know what that meant at the time). I was ruthless in my thirst for knowledge. I tried to understand power because I felt powerless. I became a TV show host, a grant maker, a faculty member, and when I finally found my calling, teaching, something strange happened. I began having extreme pain in my right shoulder blade. I saw physical therapist after physical therapist, until one of them told me: "This pain is associated with feeling like you have no community." The pain became a calling to come back home.

I had always felt the pull to return to Puerto Rico but continually told myself I was not wealthy enough, that I didn't belong to a powerful family, so why bother? I felt trapped, like I couldn't return, that I had no options. But the pain in my back wouldn't heal; I finally heeded the powerful calling and made my return to reconnect with my ancestral community. That was exactly nine months before the hurricane hit. That's how I found myself waiting in line at a gas station two days before Maria made landfall. Alone and scared.

The storm was going to enter exactly where my partner's house was located, on the southeast side of the island in a mountain town called Mariana. Luis told me, "Even if it's the worst here, I know we have community. People know where food is and how to care for each other." As someone who was raised in the second-largest city on the island, Ponce, I really didn't know what that meant—in fact, I couldn't imagine it.

On September 19, 2017, at 11 P.M., we were awakened by the sound of exploding glass. The wind and water rushed in as Luis, my cousin, and I ran to the safe room. We smashed ourselves and our pets into one tiny bathroom. We stayed there for hours. When we finally stepped out, the world had changed. The trees were stripped naked and looked like sticks; many were gone. No leaves. No sounds. A wooden chalet three doors down was completely destroyed, leaving only a retired couple who were able to survive by hiding in the concrete basement. The color had been stripped away from everything; there was barely any distinction between the wall of gray sky above us, impenetrable by sunlight, and the gray of the landscape. And the debris: one home's roof over there, a water heater down below, a window, a door. It looked like a bomb had exploded. We couldn't believe our eyes.

We were immediately displaced. We were disoriented for days, struggling to understand the depth of what had happened. We went to my cousin's house, a little farther down the road. The electricity was out, and we didn't have a gas stove. But luckily we found some rusty camping gear in the garden shed. We made a little food station in the laundry room, where we cooked what little we had.

After a few days, people in the community started coming out. They came from deeper in the mountains to start clearing the roads, cleaning the mess that was left of our things. And then we started hearing the reports—delivered largely by word of mouth.

There was only one radio station that was airing for those first few days. First we heard that only ten people had died. Then we heard that there were fridges full of unidentified bodies. (The official death toll is now 2,975.) Only those who had the obsolete landlines that belonged to the Puerto Rico telephone company had a way of communicating. No cellphones worked. They were completely useless. We were alone.

But then something shifted. During the days that followed, we realized that the neighbors became our family. Those first days, when we ran into an obstructed road, people stuck in their cars would get out as if prerehearsed, with their machetes in hand, and cut down trees to open the road. People knew what to do.

Amid supermarket rationing and lacking refrigeration or electricity, food was an ongoing challenge. On day four I wrote in my diary: "We decided we need to open up a kitchen. We are going to turn our home into a space for people to come over and eat. We need to get tables and chairs."

We had no food or water, but we did have community. We reached out to ARECMA (La Asociación Recreativa y Educativa Comunal del Barrio Mariana de Humacao), an organization that had been active in Mariana since 1982. For decades it had organized a festival as a way to encourage economic opportunity in this barrio. We asked ARECMA: Can we use your facilities? And its staff said yes. We asked our neighbors who were cooks: Would you cook? And they said yes. We asked an artist: Would you paint signs? And she said yes. We asked our friends: Would you come over and help us clean? With the little they had, they all agreed, because they knew we had to do something.

Within five days we had launched Proyecto de Apoyo Mutuo, which translates to "mutual aid project." By day ten, we were feeding three hundred people, Monday to Friday, out of completely self-organized and self-funded efforts. Donations started coming in, but the most important thing was the people on the ground, who worked tirelessly and contributed whatever food, skills, and money they had. They all contributed their whole selves so that the place was full of respect and solidarity. And the word was spreading.

People from the wider region started arriving. Pretty soon we had paramedics, nurses, and doctors. We started surveying people about what they needed. We created a distribution system that made sense for those needs. We developed a prioritization process so we could aid people who couldn't get out of their homes. And through this process of *doing*, we were able to heal instead of despair.

The way it would work is that children would come at 9:30 A.M. to clean and set up the tables. Then they would start to play chess. By 11:15 A.M. hundreds of people would arrive from Mariana and neighboring communities, all in need of food. And we would deliver meals to sixty people we identified as bedridden.

One young man who heard about what we were doing stopped us and said, "Wait, are you really helping people? Do people help anymore?" We said, "Yes! And you're invited." He showed up the next day. He told us, "I thought there weren't people like you anymore—my dad told me that good people didn't exist." That young man later decided to change his plans and stay in Puerto Rico and study on the island because he started to believe that there was a possibility for a different Puerto Rico. And that's when we looked at our work and we realized: We are building a different and better Puerto Rico.

We're building it because when everything collapsed, it was crystal clear what was the real infrastructure and what was an insubstantial facade. When everything collapses, the life-saving infrastructure is our knowledge of one another's skills, our trust of one another, our capacity to forgive our neighbor, work with our neighbor, and mobilize. When everything collapses—no ATMs, no water, no food, no diesel, no communication—you have to tap into a preexisting system of trust and dignity and reciprocity.

When disasters happen, the person right in front of you is your best chance at survival. That's when we understood: *The times we will be facing are going to require us to recognize that the most important thing around us is community.*

From that collective foundation, we can build. In Mariana we needed communication; we needed solar infrastructure; we needed ways of more sustainably sourcing water. Some six months passed before we got water service back, nine before we got power. But that didn't stop us. We braved the roads to Caguas or San Juan to access social media and send messages beyond the island. Those messages reached a network of powerful women in tech who wanted to help. Within a week of putting out a call for communication systems, we heard back from a woman who funded a company of former U.S. Coast Guard members who had responded to Hurricane Sandy. They installed Wi-Fi in Mariana within twenty-four hours. So we asked ourselves: Why can't we have free Wi-Fi for the entire community? Why can't we have our own radio station? We quickly understood that the people are the experts in knowing what they need; we just had to listen. So when women showed up with raw, purplish hands from hand-washing clothes in the river, we decided to build a solar laundromat. And we did. We have created a model of what it means for a community to be healthy enough and organized enough to face natural disasters.

Our vision now is to share markers for how we know if a community can support human adaptation to extreme weather conditions. It is so-called marginalized communities that have created alternative ways of surviving, due to the absence of basic services that are human rights. If well organized and invested in, these communities are better suited to respond when all infrastructure collapses. We owe it to the millions of people in communities like ours around the world to help them prepare for the next disaster and to live better in the meantime. We owe it to the generations that will face disproportionate burden due to climate change. We owe it to our *ancestros* who got us here. We owe it to some three thousand Puerto Ricans who died due to Hurricane Maria.

I watched as generation upon generation of false narratives collapsed. "Puerto Rico is doomed" was a false story, and the new story that emerged is completely different. This new story is being lived by the activists, organizers, and community leaders who still today are on the front lines—regardless of whether or not they can pay their bills. The truth is that our communities in Puerto Rico are powerful beyond measure. Maybe we don't make the headlines, but we are generous and collaborative. We will come up with some of the most important innovations for human adaptation in the world, because our lives depend on it.

We know why people died. It was not a natural disaster; it was a political and man-made one. Which means it can be solved. How many Boricuas would have survived if they had lived in self-organizing communities like the one I became part of after Hurricane Maria? We will never know.

The future will be challenging. All around us, the sea level will rise, and people will be displaced. People may become more and more isolated and will likely suffer from more mental health issues. The state will likely fail to protect its people—the way the United States and Puerto Rican governments have repeatedly disappointed us.

But communities like mine have always been tending to the root, telling the history erased from books, doing the most radical thing, especially for a colony: building a foundation based on dignity, abundance, and self-love. These are the examples we need to be looking toward. This is the infrastructure in need of rebuilding.

I thought I had to leave the island and all its beauty to rise on my own. And so many others like me have disengaged from their communities, overwhelmed by trying to survive, too worried about prestige and power, too jaded.

It turns out that communities are the most important force that allows humans to weather great storms, literally and metaphorically. The climate crisis will intensify, but our communities will continue to rise—because they were always standing.

ONWARD

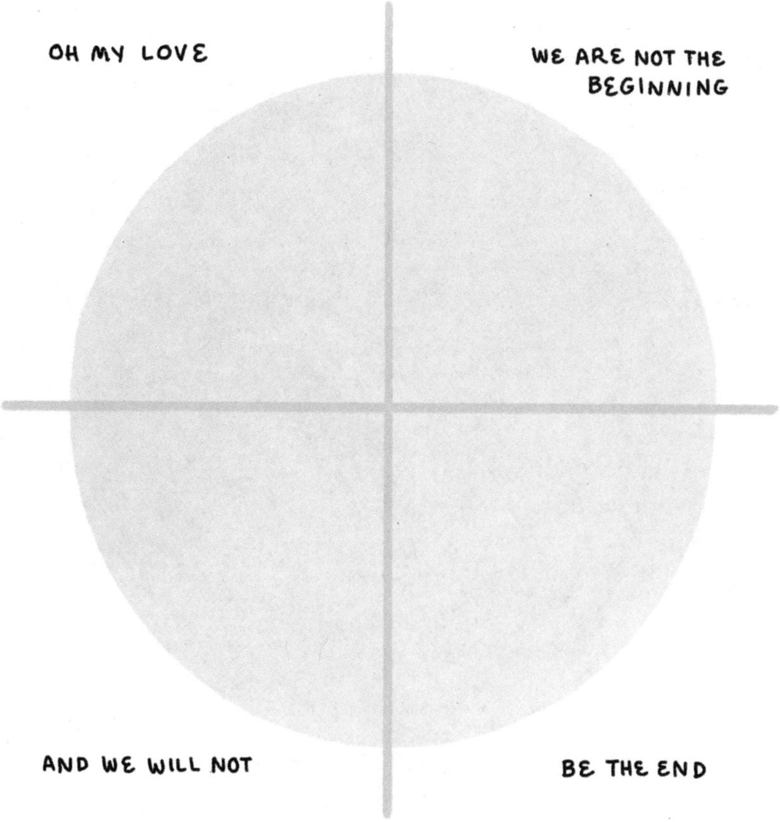

OH MY LOVE

WE ARE NOT THE
BEGINNING

AND WE WILL NOT

BE THE END

MADELEINE JUBILEE SAITO

Onward

Ayana Elizabeth Johnson and
Katharine K. Wilkinson

Nature is cyclical. It curves and revolves, with little use for linear ways. In that spirit, the end of this book is a begin-again.

We come back to the title: *All We Can Save*. Given this book's breadth and depth and choral nature, it wasn't easy to land a name that felt adequately inclusive, expansive, grounded in what is, and forward looking. Adrienne Rich's 1977 poem "Natural Resources" unlocked the title this book came to claim as its own. "My heart is moved by all I cannot save," she writes. Ours too—and by all that we can.

Each word has power: All, We, Can, Save.

"All" speaks to the whole—each and every, nothing left out. As you have read in these pages, all is connected. That's the truth both of the world's wisdom traditions and, increasingly, of scientific findings. No single thing, no species or ecosystem, community or culture is safe when so much hangs in the balance. We unravel as one or we regenerate as one. Of course, the diminished *all* we live amid and are part of today is not the same as the *all* of last year. Nor is it the same as the *all* of the 1990s, when the climate movement was born, the *all* of the late 1970s, when the fossil fuel industry knew plainly what was at risk, or the *all* of 1856, when Eunice Newton Foote identified the link between carbon dioxide and global temperature. Yet there is so much left, so much that calls for our best efforts. So "all" also speaks to the level of commitment needed: not to give up on anything, to be all in. It is naive to tether our hopes to a world without climate change and the losses it brings. It is naive to ignore how many hopes are now long shots. But that doesn't mean we quit. We must not cede the future to those who recklessly gamble it. Every tenth of a degree of warming, every centimeter of sea level rise, every increasingly unnatural disaster, every species, every life—all of it matters.

"We" speaks to the collective, to collaboration, to community, to the relational work at hand. Addressing the climate crisis, the gnarliest problem humanity has faced, will take everyone. That has to include girls and women; that has to include you; that has to include not just leaders but followers, doers, makers, and nurturers of all stripes, in true cooperation. The climate movement is now three generations deep—so much sage advice and youthful energy to harness, such diverse expertise and perspectives to bring to bear. We must hold this broad collectivism tightly to ensure poor communities, communities of color, and Indigenous peoples are not just included but at the heart of this transformation, lest we have a fractured and incomplete *we*. "We" speaks to justice, to how we do the work that needs doing and whose contributions are valued. We cannot, we must not, go it alone. To focus only on what we can do as individuals, instead of what we can do together, will mean failure. A theme that emerges strongly in this book is *community*. Indeed, building community around solutions is the most important thing.

"Can" speaks to sheer determination. This shit ain't over yet. Possibility still exists, as documented in data-driven analysis of climate solutions and temperature trajectories, and as imprinted in the persistence of life despite all odds. We are a miracle. Our task and our opportunity is to face a seemingly impossible challenge and act in service of what *is* possible. So *what can I do*? It's an increasingly prevalent question, which is a very good thing, but the answers offered are often trite, consumerist, and incomplete. And often the question should be: What can *we* do? How can we depart this perilous path? These are questions to live into every day for the rest of our lives. We can answer them (and answer them again and again) by considering what good work is unfolding around us, what invitations we may already have received, and what gifts we might have to offer, or offer more deeply. From personal acts to professional prowess to political participation, our layers of agency are more profound than we may realize. Our choices and voices, our networks, dollars, and votes, our skills and ingenuity—these are all openings for *can*. Enough of what we *cannot* achieve; *can* is the drumbeat of those who refuse to give in to destruction, who rise again and again with life force.

"Save" speaks to our opportunity and duty to protect nature, ecosystems, species, and one another. We are one another's keepers. "Save" also acknowledges how much has already been lost, and how much more loss is guaranteed to come; we are at the point of gross imbalance where "to save" is now the imperative. This word, from the Old French *sauver*—keep, protect, redeem—offers us a chance to restore and rebuild, to recommit ourselves. But saving doesn't mean holding on to or petrifying current systems—quite the opposite. It means retrieving ways of living and being that have been sidelined and suppressed. It means attending to the root causes of our enmeshed crises. It doesn't mean a draconian assignment; it means joyously building a human society that aligns with life's principles and assists the planet's living systems. As the prescient Rachel Carson put it in *Silent Spring:*

> We stand now where two roads diverge. But unlike the roads in Robert Frost's familiar poem, they are not equally fair. The road we have long been traveling is deceptively easy, a smooth superhighway on which we progress with great speed, but at its end lies disaster. The other fork of the road—the one "less traveled by"—offers our last, our only chance to reach a destination that assures the preservation of the earth.
>
> The choice, after all, is ours to make.

So where do we go from here? First, we take a breath. It's a lot. And in some ways, we, humans, were not designed for a crisis this massive and all-encompassing. In other ways, we were made for this moment. What we do now is dream. From a foundation of science and community, we must imagine the future we want to live in, and the future we want to pass on, and every day do something to reel the dream closer to reality.

Know that we already have most of the solutions we need—from regenerative farming to renewable energy to restored ecosystems to redesigned mobility, materials, and structures. We don't need to wait for new technologies or new practices; we just need to get to it, removing barriers to solutions, accelerating their implementation, and expanding their reach, while actively stopping sources of this crisis—namely fossil fuel power plants, rigs and wells, pipelines and refineries, subsidies and loopholes, as well as the destruction of ecosystems. Where national governments are not stepping up, local, city, and state actions are even more important and may hasten the federal mobilization that is vital.

Roll up your sleeves. Everyone has a role to play. And we hope, if you haven't already found yours, that through the kaleidoscope of voices and stories in this book you are starting to envision it. For there is so much that needs doing, and will continue to need doing. This is an era of transformation. This is generations of work. If no one has ever invited you to the climate movement, please consider this your warm welcome. And if no one has ever thanked you for your efforts, please hear our gratitude. You are important to this moment; you are so needed.

The work at hand is hard and uncertain, yet we find our warrior spirits, charge ahead, and care for one another every step of the way. We will stumble as we chart this unmapped path; let's forgive our fallibility, safeguard our empathy, and lead with kindness as we go. In more poignant words from Adrienne Rich:

> There must be those among whom we can sit down and weep, and still be counted as warriors.

We hope this book embodies that kindred circle. If there is one theme that runs through the collection, it is ferocious love—for one another, for Earth, for all beings, for justice, for a life-giving future. Let's move forward with love, not conquest; humility, not righteousness; generous curiosity, not hardened assumptions. It is a magnificent thing to be alive in a moment that matters so much. Let's proceed with broken-open hearts, seeking truth, summoning courage, and focused on solutions.

Gratitude

To our elders, to the women who came before, who tamped down the high grass. To the essayists in this collection and all the climate leaders who have been shaping and illuminating a path forward and welcoming so many into this work. To Madeleine Jubilee Saito, whose beautiful and wise art graces these pages. To the poets whose words bring soul, helping us feel our feelings and know we're not alone. To the One World team, especially our editor, Nicole Counts, whose enthusiasm and talent nurtured this creation, and editor-in-chief Chris Jackson, who saw what this book could be. To our literary agent, Anthony Mattero, who had our backs from the jump. To Megan Davis, eagle-eyed research assistant, and Chloe Angyal, there for edits in a pinch. To Emily Kreiger, James Gaines, and Wudan Yan for leaving no fact unturned. To the funders who graciously breathed life into this project, ensuring we could compensate all contributors well and make the book exactly what it needed to be. To our loved ones for wholeheartedly supporting the vision and us—with celebration, commiseration, snacks, patience, affection, and a willingness to listen to mediocre ideas on the way to good ones. To Louise and Lee especially. (And to Arthur dog, steadfastly at foot.) To those showing us what the future could look like, and those already building it. To those who have been doing this work for a lifetime, and those who may start now. To this magnificent planet we call home, and all life that shares it. Thank you.

Climate Solutions

"Drawdown" is the future point in time when levels of greenhouse gases in the atmosphere stop climbing and start to steadily decline. It's a critical turning point for life on this planet, and we can reach it only by stopping heat-trapping emissions *and* supporting nature's carbon sinks, which draw carbon back down to Earth. According to Project Drawdown, the world could reach drawdown by midcentury using the solutions listed here—all of which are already in hand today. They are grouped into their sectors and subgroups, rather than ranked by impact, because an entire ecosystem of solutions is critical (* indicates that a solution falls into more than one sector). This list is extensive but not exhaustive and continues to grow. For more on these climate solutions and accelerators, see *The Drawdown Review* and drawdown.org.

1. Reduce Sources—*Bringing Emissions to Zero*

Electricity

ENHANCE EFFICIENCY

- Insulation *
- High-performance glass *
- Dynamic glass *
- Green and cool roofs *
- Smart thermostats *
- Building-automation systems *
- LED lighting
- District heating *
- High-efficiency heat pumps *
- Solar hot water *
- Low-flow fixtures *
- Water-distribution efficiency

SHIFT PRODUCTION

- Concentrated solar power
- Distributed solar photovoltaics

- Utility-scale solar photovoltaics
- Micro wind turbines
- Onshore wind turbines
- Offshore wind turbines
- Geothermal power
- Small hydropower
- Ocean power
- Biomass power
- Nuclear power
- Waste-to-energy *
- Landfill methane capture *

Enhance Efficiency and Shift Production

- Building retrofitting *
- Net-zero buildings *

Improve the System

- Distributed energy storage
- Utility-scale energy storage
- Grid flexibility
- Microgrids

Food, Agriculture, and Land Use

Address Waste and Diets

- Plant-rich diets *
- Reduced food waste *

Protect Ecosystems

- Forest protection *
- Indigenous peoples' forest tenure *
- Grassland protection *
- Peatland protection and rewetting *
- Coastal wetland protection *

Shift Agriculture Practices

- Conservation agriculture *
- Regenerative annual cropping *
- Nutrient management

- Farm irrigation efficiency
- Improved rice production *
- System of rice intensification *
- Sustainable intensification for smallholders *

Industry

Improve Materials

- Alternative cement
- Bioplastics

Use Waste

- Composting
- Recycling
- Recycled paper
- Waste-to-energy *
- Landfill methane capture *
- Methane digesters *

Address Refrigerants

- Refrigerant management *
- Alternative refrigerants *

Transportation

Shift to Alternatives

- Walkable cities
- Bicycle infrastructure
- Electric bicycles
- Carpooling
- Public transit
- High-speed rail
- Telepresence

Enhance Efficiency

- Hybrid cars
- Efficient trucks
- Efficient aviation
- Efficient ocean shipping

Electrify Vehicles

- Electric cars
- Electric trains

Buildings

Enhance Efficiency

- Insulation *
- High-performance glass *
- Dynamic glass *
- Green and cool roofs *
- Smart thermostats *
- Building automation systems *
- Low-flow fixtures *

Shift Energy Sources

- District heating *
- High-efficiency heat pumps *
- Solar hot water *
- Biogas for cooking
- Improved clean cookstoves

Enhance Efficiency and Shift Energy Sources

- Building retrofitting *
- Net-zero buildings *

Address Refrigerants

- Refrigerant management *
- Alternative refrigerants *

2. Support Sinks—*Uplifting Nature's Carbon Cycle*

Land Sinks

Address Waste and Diets

- Plant-rich diets *
- Reduced food waste *

Protect and Restore Ecosystems

- Forest protection *
- Indigenous peoples' forest tenure *

- Temperate forest restoration
- Tropical forest restoration
- Grassland protection *
- Peatland protection and rewetting *

Shift Agriculture Practices

- Conservation agriculture *
- Regenerative annual cropping *
- Managed grazing
- Silvopasture
- Multistrata agroforestry
- Tree intercropping
- Perennial staple crops
- Perennial biomass production
- Improved rice production *
- System of rice intensification *
- Sustainable intensification for smallholders *

Use Degraded Land

- Abandoned farmland restoration
- Tree plantations (on degraded land)
- Bamboo production

Coastal and Ocean Sinks

Protect and Restore Ecosystems

- Coastal wetland protection *
- Coastal wetland restoration

Engineered Sinks

Remove and Store Carbon

- Biochar production

3. Improve Society—*Fostering Equality for All*

Health and Education

- Universal access to high-quality education and reproductive healthcare

Accelerators—*To Move Solutions Forward*

1. Shape culture
2. Build power
3. Set goals
4. Alter rules and policy
5. Shift capital
6. Change behavior
7. Improve technology

Referenced Organizations

Many of the essays reference climate-focused organizations that readers may want to pursue further.

B Lab
Billion Oyster Project
Biomimicry Institute
California Environmental Justice Alliance
Carbon180
Center for Cultural Power
Chattahoochee RiverLands
Citizens' Climate Lobby
Climate Action Now
Climate Justice Alliance
Climate Reality Project
CREO
Deep South Center for Environmental Justice
Dogwood Alliance
Earthjustice
Earth Uprising
Emerge Puerto Rico
Extinction Rebellion
Fire Drill Fridays
500 Women Scientists
Giniw Collective
Global Migration Project
Greenpeace
GreenWave
Gulf Coast Center for Law & Policy
Harambee House
Indigenous Environmental Network
International Living Future Institute
League of Conservation Voters
Little Village Environmental Justice Organization
Louisiana Bucket Brigade
Louisiana Environmental Action Network
Moms Clean Air Force

Montana Environmental Information Center
NAACP Environmental & Climate Justice Program
National Audubon Society
Native Renewables
Natural Resources Defense Council
NextGen America
Our Children's Trust
Project Drawdown
Protect Our Winters
PUSH Buffalo
Quivira Coalition
Re-Earth Initiative
Regeneration International
Resilience Force
RISE St. James
Sierra Club & Beyond Coal Campaign
Slow Factory Foundation
Soul Fire Farm
Sunrise Movement
Surfrider Foundation
350.org
Union of Concerned Scientists
United Farm Workers
U.S. Green Building Council
WE ACT for Environmental Justice

Select Sources

In addition to select sources shared here, you can find a full list of references online at www.allwecansave.earth/references.

Atkin, Emily. HEATED. www.heated.world.

Benyus, Janine M. *Biomimicry: Innovation Inspired by Nature*. New York: Harper Collins, 2009.

Berry, Wendell. *The Hidden Wound*. Berkeley: Counterpoint, 2010.

brown, adrienne maree. *Emergent Strategy: Shaping Change, Changing Worlds*. Chico, CA: AK Press, 2017.

Bullard, Robert, and Beverly Wright. *The Wrong Complexion for Protection: How the Government Response to Disaster Endangers African American Communities*. New York: New York University Press, 2012.

Butler, Octavia E. *Parable of the Sower*. New York: Warner Books, 1993.

Carson, Rachel. *Silent Spring*. New York: Houghton Mifflin, 1962.

Chenoweth, Erica. "The Success of Nonviolent Civil Resistance." Filmed November 4, 2013, at TEDxBoulder, Boulder, CO. Video, 12:33.

Collins, Patricia Hill. *Black Feminist Thought: Knowledge, Consciousness, and the Politics of Empowerment*. New York: Routledge, 2000.

"A Feminist Agenda for a Green New Deal: Principles and Values." No date. www.feministgreennewdeal.com/principles.

Foote, Eunice. "Circumstances Affecting the Heat of the Sun's Rays." *American Journal of Science and Arts* 22 (1856): 382–83.

Goodell, Jeff. *The Water Will Come: Rising Seas, Sinking Cities, and the Remaking of the Civilized World*. New York: Little, Brown, 2017.

Green America and Kiss the Ground. "Climate Victory Gardens." No date. www.greenamerica.org/climate-victory-gardens.

Hayhoe, Katharine. "The Most Important Thing You Can Do to Fight Climate Change: Talk About It." Filmed November 30, 2018, at TED Women, Palm Springs, CA. Video, 17:04.

Intergovernmental Panel on Climate Change. "Global Warming of 1.5°C." Edited by Valérie Masson-Delmotte et al. Special report, 2019. www.ipcc.ch/sr15.

Kimmerer, Robin Wall. *Braiding Sweetgrass: Indigenous Wisdom, Scientific Knowledge, and the Teachings of Plants*. Minneapolis: Milkweed Editions, 2013.

Klein, Naomi. *On Fire: The (Burning) Case for a Green New Deal*. New York: Simon & Schuster, 2019.

Kolbert, Elizabeth. *The Sixth Extinction: An Unnatural History*. New York: Henry Holt, 2014.

Leiserowitz, Anthony, et al. *Climate Change in the American Mind*. New Haven: Yale Program on Climate Change Communication, 2019.

Luxemburg, Rosa. *The Accumulation of Capital*. New York: Routledge & Kegan Paul, 1951.

Macy, Joanna, and Chris Johnstone. *Active Hope: How to Face the Mess We're In Without Going Crazy*. Novato, CA: New World Library, 2012.

Mitchell, Sherri. *Sacred Instructions: Indigenous Wisdom for Living Spirit-Based Change*. Berkeley, CA: North Atlantic Books, 2018.

Moms Clean Air Force. "Breath of Life: Bible Study Curriculum." No date.

Ocasio-Cortez, Alexandria, and Avi Lewis. "A Message from the Future with Alexandria Ocasio-Cortez." Illustrated by Molly Crabapple. Presented by Naomi Klein and The Intercept, April 17, 2019. Video, 7:35.

Oreskes, Naomi, and Eric M. Conway. *Merchants of Doubt: How a Handful of Scientists Obscured the Truth on Issues from Tobacco Smoke to Global Warming*. New York: Bloomsbury, 2010.

Penniman, Leah. *Farming While Black: Soul Fire Farm's Practical Guide to Liberation on the Land*. White River Junction, VT: Chelsea Green, 2018.

Project Drawdown. *The Drawdown Review: Climate Solutions for a New Decade*. Edited by Katharine Wilkinson. San Francisco: Project Drawdown, 2020.

Ray, Janisse. *The Seed Underground: A Growing Revolution to Save Food*. White River Junction, VT: Chelsea Green, 2012.

Robinson, Mary. *Climate Justice: Hope, Resilience, and the Fight for a Sustainable Future*. New York: Bloomsbury, 2018.

Robinson, Mary, and Maeve Higgins. *Mothers of Invention* (podcast). Produced by Doc Society. www.mothersofinvention.online.

Schwartz, Judith D. *Water in Plain Sight: Hope for a Thirsty World*. White River Junction, VT: Chelsea Green, 2019.

Sellers, Sam. "Gender and Climate Change in the United States: A Reading of Existing Research." Women's Environment and Development Organization and Sierra Club, 2020.

Shelley, Mary. *Frankenstein; or, The Modern Prometheus*. London: Lackington, Hughes, Harding, Mavor, & Jones, 1818.

Simard, Suzanne. "How Trees Talk to Each Other." Filmed June 29, 2016, at TEDSummit, Banff, Canada. Video, 18:11.

Smith, Bren. *Eat Like a Fish: My Adventures as a Fisherman Turned Restorative Ocean Farmer*. New York: Alfred A. Knopf, 2019.

Southwest Network for Environmental and Economic Justice. "*Jemez Principles* for Democratic Organizing." 1996. www.climatejusticealliance.org/jemez-principles.

Stokes, Leah Cardamore. *Short Circuiting Policy: Interest Groups and the Battle over Clean Energy and Climate Policy in the American States*. New York: Oxford University Press, 2020.

Sturgeon, Amanda. *Creating Biophilic Buildings*. Seattle: International Living Future Institute, 2017.

Thunberg, Greta. *No One Is Too Small to Make a Difference*. New York: Penguin Books, 2019.

Toensmeier, Eric. *The Carbon Farming Solution: A Global Toolkit of Perennial Crops and Regenerative Agriculture Practices for Climate Change Mitigation and Food Security*. White River Junction, VT: Chelsea Green, 2016.

U.S. Congress. Recognizing the Duty of the Federal Government to Create a Green New Deal, H.R. 109, 116th Cong., § 1 (2019).

United States Global Change Research Program. "Fourth National Climate Assessment." 2017 and 2018. nca2018.globalchange.gov.

Warren, Elizabeth. "Climate Plans." elizabethwarren.com/climate.

Westervelt, Amy. *Drilled* (podcast). Produced by Critical Frequency. criticalfrequency.org/drilled.

Williams, Terry Tempest. *Erosion: Essays of Undoing*. New York: Farrar, Straus and Giroux, 2019.

Poetry

Bass, Ellen. *The Human Line*. Port Townsend, WA: Copper Canyon, 2012.

Dungy, Camille T. *Trophic Cascade*. Middletown, CT: Wesleyan University Press, 2017.

Fisher-Wirth, Ann, and Laura-Gray Street, eds. *The Ecopoetry Anthology*. San Antonio: Trinity University Press, 2013.

Garcia, Alixa, and Naima Penniman. *Climbing Poetree*. Seattle: Whit, 2014.

Harjo, Joy. *An American Sunrise: Poems*. New York: W. W. Norton, 2019.

Hirshfield, Jane. *Ledger: Poems*. New York: Alfred A. Knopf, 2020.

Hopper, Ailish. *Dark~Sky Society*. Kalamazoo, MI: New Issues, 2014.

Kane, Joan Naviyuk. *Another Bright Departure*. Missoula, MT: Cutbank Books, 2018.

Limón, Ada. *The Carrying: Poems*. Minneapolis: Milkweed Editions, 2018.

Olds, Sharon. *Odes*. New York: Alfred A. Knopf, 2016.

Oliver, Mary. *Devotions: The Selected Poems of Mary Oliver*. New York: Penguin Press, 2017.

Pierce, Catherine. *The Tornado Is the World*. Philadelphia: Saturnalia Books, 2016.

Piercy, Marge. *Circles on the Water: Selected Poems of Marge Piercy*. New York: Alfred A. Knopf, 2009.

Rich, Adrienne. *The Dream of a Common Language: Poems 1974–1977*. New York: W. W. Norton, 1978.

Rilke, Rainer Maria. *The Selected Poetry of Rainer Maria Rilke*. Edited by Stephen Mitchell. New York: Vintage International, 1989.

Smith, Patricia. *Blood Dazzler: Poems*. Minneapolis: Coffee House, 2008.

Tuckey, Melissa, ed. *Ghost Fishing: An Eco-justice Poetry Anthology*. Athens: University of Georgia Press, 2018.

Walker, Alice. *Hard Times Require Furious Dancing*. Novato, CA: New World Library, 2010.

Credits

Grateful acknowledgment is made to the following for permission to print previously published material:

Bass, Ellen: "The Big Picture" from THE HUMAN LINE, copyright © 2007 by Ellen Bass. Used by permission of The Permissions Company, LLC, on behalf of Copper Canyon Press, coppercanyonpress.org. • Battle, Colette Pichon: "An Offering from the Bayou" is evolved from the following TED Talk: "Climate Change Will Displace Millions. Here's How We Prepare," filmed at TEDWomen in December 2019. Used by permission of the author and TED. • brown, adrienne maree: "What Is Emergent Strategy?" from EMERGENT STRATEGY: SHAPING CHANGE, CHANGING WORLDS, copyright © 2017. Used by permission of AK Press. • Dungy, Camille T.: "Characteristics of Life" from TROPHIC CASCADE, copyright © 2017 by Camille T. Dungy, published by Wesleyan University Press, Middletown, CT, and used by permission of the publisher. • Harjo, Joy: "For Those Who Would Govern" from AN AMERICAN SUNRISE: POEMS, copyright © 2019 by Joy Harjo. Used by permission of W. W. Norton & Company, Inc. • Hirshfield, Jane: "On the Fifth Day" from LEDGER: POEMS, copyright © 2020 by Jane Hirshfield. Used by permission of Alfred A. Knopf, an imprint of the Knopf Doubleday Publishing Group, a division of Penguin Random House LLC. All rights reserved. • Hopper, Ailish: "Did It Ever Occur to You That Maybe You're Falling in Love" originally published in *Poetry*, January 2016, vol. 207, no. 4. Used by permission of the author. • Kane, Joan Naviyuk: "The Straits" from MILK BLACK CARBON, copyright © 2017. Used by permission of the University of Pittsburgh Press. • Limón, Ada: "Dead Stars" from THE CARRYING (Minneapolis: Milkweed Editions, 2018). Copyright © 2018 by Ada Limón. Used by permission of Milkweed Editions, Milkweed.org. • Olds, Sharon: "Ode to Dirt" from ODES, compilation copyright © 2016 by Sharon Olds. Used by permission of Alfred A. Knopf, an imprint of the Knopf Doubleday Publishing Group, a division of Penguin Random House LLC. All rights reserved. • Oliver, Mary: "Mornings at Blackwater" from RED BIRD, published by Beacon Press, Boston, copyright © 2008 by Mary Oliver. Used by permission of the Charlotte Sheedy Literary Agency, Inc. • Penniman, Naima: "Being Human" from CLIMBING POETREE by Alixa Garcia and Naima Penniman, copyright © 2014 by Climbing Poetree. Used by permission of Whit Press. • Pierce, Catherine: "Anthropocene Pastoral" originally published in *American Poetry Review*, vol. 46, no. 06. Collected in DANGER DAYS (2020). Used by permission of the author. • Piercy, Marge: "To Be of Use" from CIRCLES ON THE WATER, copyright © 1982 by Middlemarsh, Inc. Used by permission of Alfred A. Knopf, an imprint of the Knopf Doubleday Publishing Group, a division of Penguin Random House LLC. All rights reserved. • Prakash, Varshini: Some passages within "We Are Sunrise" were taken from "Generation Climate Change: Ezra Sits Down with Varshini Prakash of Sunrise Movement," *The Ezra Klein Show*, July 31, 2019. Used by permission of the author and VOX. • Ray, Janisse: Excerpt from THE SEED UNDERGROUND, copyright © 2012 by Janisse Ray. Used by permission of Chelsea Green Publishing, www.chelseagreen.com. • Rich, Adrienne: "Natural Resources" from COLLECTED POEMS: 1950–2012, copyright © 2016, 2013 by the Adrienne Rich Literary Trust. Copyright © 1978 by W. W. Norton & Company, Inc. Used by permission of W. W. Norton & Company, Inc. • Rodriguez, Christine E. Nieves: "Community Is Our Best Chance" is evolved from the following TED Talk: "Why Community Is Our Best Chance for Survival—a Lesson Post–Hurricane Maria." Filmed at TEDMED in November 2018. Used by permission of the author and TED. • Smith, Patricia: "Man on the TV Say" from BLOOD DAZZLER, copyright © 2008 by Patricia Smith. Used by permission of The Permissions Company, LLC on behalf of Coffee House Press, coffeehousepress.org. • Walker, Alice: "Calling All Grand Mothers" from HARD TIMES REQUIRE FURIOUS DANCING, copyright © 2010 by Alice Walker. Used by permission of New World Library, Novato, CA, www.newworldlibrary.com.

Some of the essays in this collection are iterations of or include excerpts from previous publications:

Atkin, Emily: "Truth Be Told" originally appeared in different form as "Good Grief" in *Columbia Journalism Review*, Spring 2020. Used by permission of the author. • Benyus, Janine: "Reciprocity" originally published in different form under the same title in DRAWDOWN: THE MOST COMPREHENSIVE PLAN EVER PROPOSED TO REVERSE GLOBAL WARMING by Project Drawdown, published by Penguin Books, 2017. Used by permission of the author. • Gunn-Wright, Rhiana: "A Green New Deal for All of Us" is an adapted version of "Policies and Principles of a Green New Deal" from WINNING THE GREEN NEW DEAL: WHY WE MUST, HOW WE CAN, edited by Varshini Prakash and Guido Girgenti, published by Simon & Schuster, 2020. Used by permission of the author. • Hayhoe, Katharine: "How to Talk About Climate Change So People Will Listen" originally appeared in different form in *Chatelaine*, April 18, 2019. Used by permission of the author. • Heglar, Mary Annaïse: "Home Is Always Worth

It" originally appeared in different form on *Medium,* September 12, 2019. Used by permission of the author. • Johnson, Ayana Elizabeth: Some material in "Onward" was originally published in "There Is Nothing Naive About Moral Clarity," originally published on *Medium,* September 27, 2019. Used by permission of the author. • Johnston, Emily N.: "Loving a Vanishing World" originally published in different form on *Medium,* May 9, 2019. Used by permission of the author. • Klein, Naomi: "On Fire" originally appeared in slightly different form in ON FIRE: THE (BURNING) CASE FOR A GREEN NEW DEAL, published by Simon & Schuster, 2019. Used by permission of the author. • Marvel, Kate: Some material in "A Handful of Dust" was originally published in "This Was the Decade We Knew We Were Right," *Scientific American,* December 30, 2019. Used by permission of the author. • Miller, Sarah: "Heaven or High Water" was originally published in different form in *Popula,* April 2, 2019. Used by permission of the author. • Penniman, Leah: Some material in "Black Gold" originally appeared in "By Reconnecting with Soil, We Heal the Planet and Ourselves," *Yes!* February 14, 2019, and "Black Farmers Embrace Practices of Climate Resiliency," *Yes!* December 18, 2019. Used by permission of the author. • Sanders, Ash: "Under the Weather" was originally published in different form in *The Believer,* December 2, 2019. Used by permission of the author. • Schwartz, Judith D.: Some material in "Water Is a Verb" originally appeared in WATER IN PLAIN SIGHT: HOPE FOR A THIRSTY WORLD, published by Chelsea Green, 2019. Used by permission of the author. • Toney, Heather McTeer: Some material in "Collards Are Just As Good As Kale" was originally published in "Black Women Are Leaders in the Climate Movement," *New York Times,* July 25, 2019. Used by permission of the author. • Westervelt, Amy: "Mothering in an Age of Extinction" was originally published in different form in *Drilled News,* December 19, 2019. Used by permission of the author. • Wilkinson, Katharine: Some material in "Begin" was originally published in "The Woman Who Discovered the Cause of Global Warming Was Long Overlooked. Her Story Is a Reminder to Champion All Women Leading on Climate," *Time,* July 17, 2019, and "Women, Girls, and Nonbinary Leaders Are Demonstrating the Kind of Leadership Our World So Badly Needs," *The Elders,* December 6, 2019. Used by permission of the author.

Index

Note: Page numbers in **bold** refer to contributed essays.

About the Contributors

Emily Atkin is an award-winning environmental reporter and the author and founder of HEATED, a daily climate change newsletter. She is also a contributing columnist at MSNBC.

Ellen Bass's latest book is *Indigo*. Her poems appear frequently in *The New Yorker*. A chancellor of the Academy of American Poets, she teaches in the MFA program at Pacific University.

Xiye Bastida is a climate activist who has brought the voices of Indigenous peoples and youth to the forefront of climate solutions. She was raised in Mexico and attends the University of Pennsylvania.

Colette Pichon Battle, JD, is a human rights attorney, nonprofit executive, and daughter of the bayou. She is an expert on climate disaster impacts and coarchitect of the Gulf South for a Green New Deal.

Jainey K. Bavishi leads New York City's efforts to prepare for the impacts of climate change. She has previously worked in New Orleans, Louisiana, and Honolulu, Hawaii.

Janine Benyus is a biologist, innovation consultant, and author of *Biomimicry*, the book that sparked the nature-inspired design movement. She is cofounder of Biomimicry 3.8 and the Biomimicry Institute.

adrienne maree brown is the author of *Pleasure Activism* and *Emergent Strategy* and the co-editor of *Octavia's Brood.* She is the co-host of two podcasts, *How to Survive the End of the World* and *Octavia's Parables.*

Régine Clément is a Brooklyn-based entrepreneur focused on advancing social and environmental causes and CEO of CREO, an organization focused on catalyzing capital into sustainability solutions.

Abigail Dillen, JD, is the president of Earthjustice, which represents more than five hundred clients free of charge, harnessing the power of law to force climate solutions while protecting healthy communities and ecosystems.

Camille T. Dungy is a poet, scholar, essayist, and the author of four collections of poetry, most recently *Trophic Cascade*, winner of the Colorado Book Award. She was awarded a Guggenheim Fellowship in 2019.

Rhiana Gunn-Wright is a policy expert and Chicago native who helped to develop the Green New Deal. She is the director of climate policy at the Roosevelt Institute and a 2013 Rhodes Scholar.

Joy Harjo is an internationally renowned performer and writer of the Muscogee (Creek) Nation and was named the 23rd Poet Laureate of the United States in 2019. Her most recent collection is *An American Sunrise*.

Katharine Hayhoe, PhD, is a climate scientist and UN Champion of the Earth who can often be found talking to people about why climate change matters to us and what we can do to fix it.

Mary Annaïse Heglar is a writer, communications professional, and podcast host based in New York City. Her work explores the connections among the climate crisis, justice, and emotionality.

Jane Hirshfield's ninth poetry collection, *Ledger*, faces into the crises of biosphere and social contract. She was elected into the American Academy of Arts and Sciences in 2019.

Mary Anne Hitt is the national director of campaigns at Sierra Club, former director of the Beyond Coal Campaign, and co-host of the podcast No Place Like Home. She lives in West Virginia.

Ailish Hopper is a poet, multidisciplinary artist, teacher, and writer. Her most recent book of poems is *Dark-Sky Society*.

Tara Houska—Zhaabowekwe, JD, is Couchiching First Nation Ojibwe, an attorney, environmental and Indigenous rights advocate, and founder of Giniw Collective. She lives in a pipeline resistance camp in Minnesota.

Emily N. Johnston is a poet, essayist, and cofounder of 350 Seattle. Her book, *Her Animals*, was a finalist for the 2016 Washington State Book Award.

Joan Naviyuk Kane is the author of seven books and chapbooks of poetry and prose, most recently *Another Bright Departure*. She received a 2018 Guggenheim Fellowship and a 2014 American Book Award.

Naomi Klein is an award-winning journalist and the bestselling author of books including *The Shock Doctrine* and *This Changes Everything*. She is a senior correspondent for The Intercept and a professor at Rutgers University.

Kate Knuth, PhD, is a climate citizen and transformation scholar. She is the founder of Democracy and Climate LLC and previously served as a member of the Minnesota House of Representatives.

Ada Limón, a current Guggenheim fellow, is the author of five books of poetry, including *The Carrying*, which won the National Book Critics Circle Award for Poetry.

Louise Maher-Johnson is a regenerative farmer who daily observes the intelligence and beauty in nature. She smiles at the prowess and antics of the heritage laying hens as they meander through her Climate Victory Garden.

Kate Marvel, PhD, is a climate scientist and writer living in New York City. She has a doctorate in astrophysics, so she knows Earth is the best place in the entire universe.

Gina McCarthy is the White House National Climate Advisor. Formerly, she was president and CEO of NRDC and administrator of the U.S. Environmental Protection Agency.

Anne Haven McDonnell teaches creative writing and climate justice at the Institute of American Indian Arts in Santa Fe, New Mexico. Her most recent publication is the chapbook *Living with Wolves*.

Sarah Miller is a writer who lives in Nevada County, California. You can find her work in *The New York Times*, *The Cut*, Popula.com, NewYorker.com, *The Outline*, and *Commune*.

Sherri Mitchell—Weh'naHa'mu Kwasset, JD, is a Native American attorney, teacher, activist, and change maker. She is the author of *Sacred Instructions: Indigenous Wisdom for Living Spirit-Based Change*.

Susanne C. Moser, PhD, is a social science researcher, consultant, writer, and speaker. A leader in the U.S. climate adaptation field, she directs her own research and consulting firm from western Massachusetts.

Lynna Odel is an environmental engineer who loves words. She believes climate justice is a spiritual issue.

Sharon Olds is the author of numerous books of poetry. Among her awards, *Stag's Leap* won the Pulitzer Prize and *The Dead and the Living* won the National Book Critics Circle Award.

Mary Oliver was a prolific American poet. Her honors include fellowships from the Guggenheim Foundation and the National Endowment for the Arts, a Pulitzer Prize, and a National Book Award.

Kate Orff is a designer focused on climate adaptation and biodiversity in the built environment. She is the founder of the landscape architecture practice SCAPE and a professor at Columbia University.

Jacqui Patterson is a researcher, policy analyst, organizer, and activist. She is the senior director of the NAACP Environmental and Climate Justice Program and cofounder of Women of Color United.

Leah Penniman is a Black Kreyol farmer, author, and food justice activist who founded Soul Fire Farm with the mission to end racism in the food system and reclaim our ancestral connection to land.

Naima Penniman is a multi-dimensional artist, activist, healer, grower, and educator committed to planetary health and community resilience. She is the co-founder and co–artistic director of CLIMBING POETREE.

Catherine Pierce is a poet and professor of English at Mississippi State University. An NEA fellow and Pushcart Prize winner, Pierce is the author of four books of poetry, most recently *Danger Days*.

Marge Piercy is the author of nineteen volumes of poetry and seventeen novels, including the *New York Times* bestseller *Gone to Soldiers*. Her newest book of poetry is *On the Way Out, Turn Off the Light*.

Kendra Pierre-Louis is a climate reporter with the podcast *How to Save a Planet*. She has worked for *The New York Times*, *Popular Science*, and InsideClimate News, and authored the book *Green Washed*.

Varshini Prakash is executive director and cofounder of Sunrise, a movement of young people working to stop climate change and build economic prosperity for all through a just and ambitious Green New Deal.

Janisse Ray is a writer, naturalist, activist, and seed-saver. She is the author of several books, including *The Seed Underground*, *Pinhook*, and *Ecology of a Cracker Childhood*, a *New York Times* Notable Book.

Christine E. Nieves Rodriguez is a mother, speaker, writer, and emergent strategy practitioner. She is cofounder and executive director of Emerge Puerto Rico, a climate change leadership organization.

Favianna Rodriguez is an award-winning artist, cultural strategist, and president of the Center for Cultural Power, a national organization investing in artists as agents of social change.

Cameron Russell is a model, writer, and organizer. She is the cofounder of the Model Mafia. Her work leverages creative collaboration and collective storytelling to facilitate evolution.

Madeleine Jubilee Saito is a cartoonist and designer living on Massachusett and Wampanoag land in Somerville, Massachusetts. She is creative director of The All We Can Save Project and makes work about climate justice and the sacred.

Ash Sanders is a writer, audio producer, and climate activist. She's a proud Utahan currently residing in Brooklyn.

Judith D. Schwartz is a Vermont-based author and journalist who writes and speaks about nature-based solutions to global environmental challenges. Her latest book is *The Reindeer Chronicles*.

Patricia Smith is the author of eight books of poetry, including *Incendiary Art*, winner of the Kingsley Tufts Poetry Award, the *Los Angeles Times* Book Prize, and finalist for the Pulitzer Prize.

Emily Stengel is cofounder and co–executive director of GreenWave, a nonprofit that trains and supports regenerative ocean farmers in the era of climate change.

Sarah Stillman is a staff writer for *The New Yorker* and launched the Global Migration Project at Columbia Journalism School. She teaches writing at Yale University and is a MacArthur Fellow.

Leah Cardamore Stokes, PhD, is a professor at the University of California at Santa Barbara and a policy expert on energy and climate change. She is the author of *Short Circuiting Policy*.

Amanda Sturgeon is an award-winning architect for the firm Mott MacDonald, where she brings to life buildings that connect people and nature. She is the author of *Creating Biophilic Buildings*.

Maggie Thomas is chief of staff for the White House Domestic Climate Policy Office. Previously, she served as climate policy adviser to the presidential campaigns of Gov. Jay Inslee and Sen. Elizabeth Warren.

Heather McTeer Toney is a Mississippi Delta native, climate justice liaison at EDF, senior advisor to Moms Clean Air Force, and former EPA regional administrator. She was the first African American mayor of Greenville, Mississippi.

Alexandria Villaseñor is a teenage climate activist, community organizer, and public speaker from New York City. She is also the founder and executive director of Earth Uprising.

Alice Walker is a celebrated writer, poet, and activist whose books include seven novels and volumes of essays and poetry. She won the Pulitzer Prize for Fiction and the National Book Award in 1983.

Amy Westervelt is an award-winning climate journalist, founder of the Critical Frequency podcast network, reporter/host/producer of the climate podcast *Drilled*, and co-host of the podcast and newsletter *Hot Take.*

Jane Zelikova, PhD, is an ecologist working at the intersection of climate change science, policy, and communication. She is the cofounder of 500 Women Scientists and a research scientist at the University of Wyoming.

About the Editors

Dr. Ayana Elizabeth Johnson is a marine biologist, policy expert, writer, and Brooklyn native. She is co-founder of Urban Ocean Lab, a think tank for the future of coastal cities, and co-creator of *How to Save a Planet*, a podcast on climate solutions. With Dr. Katharine Wilkinson, she co-founded The All We Can Save Project. Dr. Johnson co-created the Blue New Deal, a road map for including the ocean in climate policy. Previously, she was executive director of the Waitt Institute, developed policy at the EPA and NOAA, served as a leader of the March for Science, and taught at New York University. Dr. Johnson earned a BA from Harvard University in environmental science and public policy and a PhD from Scripps Institution of Oceanography in marine biology. *Elle* named her one of "27 women leading the charge to protect our environment," *Outside* magazine called her "the most influential marine biologist of our time," and she is on the 2021 *Time*100 Next list. Her mission is to build community around solutions to our climate crisis. She can be found online at ayanaelizabeth.com and on social media @ayanaeliza.

Dr. Katharine K. Wilkinson is an author, strategist, and teacher working to heal the planet we call home. Her writings include *The Drawdown Review* (2020), the *New York Times* bestseller *Drawdown* (2017), and *Between God & Green* (2012), which *The Boston Globe* dubbed "a vitally important, even subversive, story." Dr. Wilkinson co-founded and leads The All We Can Save Project with Dr. Ayana Elizabeth Johnson, and co-hosts the podcast *A Matter of Degrees* with Dr. Leah Stokes. Previously, she was the principal writer and editor-in-chief at the climate solutions nonprofit Project Drawdown. Dr. Wilkinson speaks widely, including a TED talk on climate and gender equality with more than 1.9 million views. A homegrown Atlantan, she holds a DPhil in geography and environment from the University of Oxford, where she was a Rhodes Scholar, and a BA in religion from Sewanee: The University of the South. Formative months spent in the Southern Appalachians, as a student at the Outdoor Academy, shaped her path. *Time* magazine featured Dr. Wilkinson as one of fifteen "women who will save the world," and Apolitical named her one of the "100 most influential people in gender policy." She's happiest on a mountain or a horse. She can be found online at kkwilkinson.com and on social media @drkwilkinson.

www.allwecansave.earth • @allwecansave

PHOTO: © JENNIFER ROBINSON